Computer-Aided Figured Graph Designing for Jacquard Weaving

Computer-aided figured graph designing (CAFGD) has become essential for the electronic jacquard weaving sector. This book explains the common principles of CAFGD for jacquard weaving, including systematic fundamental common procedure for doing computer-aided graph designing for jacquard weaving using any software. It covers the mathematical calculations to be followed in different stages of converting the given motif to figured graph design with required weave inputs. It also covers the common algorithms followed in preparing computer-aided figured graph designing.

Features:

- Talks about the application or use of computers or digital platforms to design the figured graph designing for jacquard weaving.
- Covers the mathematical calculations to be followed in different stages of converting the given motif to figured graph design with required weave inputs.
- Discusses principles of computer-aided figured graph designing for jacquard weaving.
- Includes weave creation and weave mapping as per the variety of jacquard fabric.
- Reviews transfer of digital graph for card punching used in mechanical jacquard or directly to electronic jacquard.

This book is aimed at graduate students and researchers in textile engineering, textiles, and fashion design.

Computer-Aided Figured Graph Designing for Jacquard Weaving

R. G. Panneerselvam, C. Prakash, and
P. Nagabhushanam

CRC Press
Taylor & Francis Group
Boca Raton London New York

CRC Press is an imprint of the
Taylor & Francis Group, an **informa** business

Designed cover image: Authors

First edition published 2025
by CRC Press
2385 NW Executive Center Drive, Suite 320, Boca Raton FL 33431

and by CRC Press
4 Park Square, Milton Park, Abingdon, Oxon, OX14 4RN

CRC Press is an imprint of Taylor & Francis Group, LLC

ISBN: 978-1-032-45288-3 (hbk)
ISBN: 978-1-032-57829-3 (pbk)
ISBN: 978-1-003-44120-5 (ebk)

DOI: 10.1201/9781003441205

Typeset in Times
by codeMantra

Dedicated to all the Teachers of the Indian Institute of Handloom Technology, Salem, from whom we learned all about fabric structures, figured graph designing, and jacquard weaving.

Contents

About the Authors

R. G. Panneerselvam is currently working as an Associate Professor with the School of Fashion Technology at KCG College of Technology, Karapakkam, Chennai. After serving for two decades in the profession of teaching at the Indian Institute of Handloom Technology (IIHT), Salem, and for 10 years in R&D activities at Weavers' Service Centres (WSC), functioning under the Development Commissioner (Handlooms), New Delhi, he retired as the Director of IIHT, Varanasi. His teaching interests include computer-aided jacquard graph designing; woven fabric structures and analysis; manual designing for heald, dobby, and jacquard shedding; and woven fabric manufacture and calculations. His research areas include the development of orthogonal structures for weaving face-back-face fabrics, the use of different editing software for jacquard graph designing, and the dissemination of ergonomic weaving techniques for handloom weavers for weaving sustainable handloom fabrics. He has completed his Ph.D. in Handloom and Textiles at Gandhigram Rural Institute, Gandhigram, Dindigul, Tamil Nadu. He has been actively engaging himself in product diversification and process modernization for the handloom sector and personal training for the handloom weavers. He has published about 11 peer-reviewed papers in national and international journals and the book *Weaving of Face-Flip-Face Fabric Using Orthogonal Weft Tapestry*. He is a gold medallist for securing the first rank in the Diploma and also in the AMIE examination. The Institution of Engineers (India) felicitated him as an 'Eminent Engineering Personality', recognizing his valuable contribution to Textile Engineering.

C. Prakash is Director at the Indian Institute of Handloom Technology (IIHT), Fulia, West Bengal. He also worked as a teaching faculty in engineering colleges. He obtained his Ph.D. in 2014 from Anna University, Chennai, and holds B.E. and M.Tech. degrees in Textile Technology from Pavendar Bharathidasan College, Trichy, and Kumaraguru College of Technology, Coimbatore. He has around 20 years of experience in teaching and industry. In total, he has published 200+ technical papers in reputed SCI, Scopus journals, and conferences. He is highly interested in academic and research activities in the areas concerning thermal comfort, moisture management of textile products, and effluent treatments.

P. Nagabhushanam is a retired faculty member from the Indian Institute of Handloom Technology (IIHT), Govt. of India, Ministry of Textiles, Salem, Tamil Nadu. He is currently working as a Visiting Faculty at the National Institute of Design (NID), Andhra Pradesh, and conducting classes in Basic Weaving, Advanced Weaving, Loom Mechanism, and Textile Mathematics. His teaching interests include computer-aided jacquard graph designing, woven fabric structures and analysis, and designing for heald, dobby, and jacquard shedding. He is also an expert in woven fabric manufacturing, cloth calculations, the development of new weave structures for weaving reversible fabrics, and ergonomic weaving techniques to weave sustainable handloom fabrics. He has completed the Diploma in Handloom Technology at IIHT, Salem, Tamil Nadu. He had been in the teaching profession for three decades at IIHT, Salem, and in a Special Rehabilitation Project for 5 years at the A. P. State Textile Development Corporation, Hyderabad, to provide continuous employment and rehabilitate the handloom weavers affected by the cyclone and tidal waves of 1977. He has published two peer-reviewed papers in journals.

Foreword

At the outset, I am happy to know that the team of past and present faculty of the Indian Institute of Handloom Technologies (IIHTs) has come out with a book on 'Computer-Aided Figured Graph Designing for Jacquard Weaving'. The six IIHTs functioning under the office of the Development Commissioner for Handlooms (DCH), Ministry of Textiles, Government of India, cater to the technological upgradation in handlooms and train the technical personnel to serve the handloom industry by offering the Diploma and UG programs in Handlooms and Textiles. The curriculum of these programs, apart from other courses, includes courses on structures, technology, and analysis of woven fabrics together with figured graph designing for jacquard weaving.

When electronic jacquards started replacing mechanical jacquards in the weaving sectors of the textile industry, learning 'Computer-Aided Figured Graph Designing (CAFGD)', in addition to manual graph designing, became essential to preparing digital figured graphs using any CAFGD software for electronic jacquard weaving. The 'Computer-Aided Figured Graph Designing (CAFGD)' course was introduced in the IIHT curriculum in this context. With their knowledge gained in teaching the CAFGD in IIHT for two decades, the authors have designed this book as a practical guide for beginners in CAFGD, covering all the aspects of figured graph designing. This book explains the common principles of CAFGD for jacquard weaving. After going through this book, the designers can understand the tools and options of any CAFGD software and carry out the digital jacquard graph designing with much more confidence.

This book also summarizes how the CAFGD introduced for preparing digital graphs for the electronic jacquard has also become useful to punch cards for the mechanical jacquard used in the power loom and handloom weaving sectors. Hence, this book, apart from being a textbook for graduates in different fields of the textile industry, will be a practical guide for the designers, researchers, and technocrats engaged with figured woven fabric production in the organized weaving sector using the electronic jacquard, the decentralized power loom, and cottage handloom sectors using the mechanical jacquard.

It gives me immense pleasure that this book is being released at an appropriate time when the handloom industry started using intermittently operated electronic jacquard in silk handloom clusters to weave contemporary high-end silk sarees with multicolor warp and weft traditional tapestry designs for the niche market. I compliment the authors on their endeavours and hope that they will contribute further to the literature on handloom technology.

Dr. M. Beena IAS
Development Commissioner (Handlooms)
MINISTRY OF TEXTILES
GOVERNMENT OF INDIA
Room No. 56, Udyog Bhawan,
New Delhi - 110011
Tel: 011-2306 2945, 2306 3684,
E-mail: dchl@nic.in,
Website: handlooms.nic.in

Preface

In the textile industry, different woven fabric structures are produced using healds, dobby, and jacquard shedding devices. Heald and dobby produce the structures in all-over, brick, and simple-figured motif styles. Elaborate figured motif structures are produced using the jacquard. Diversified figured styles of compound structures are produced using the huge hook capacity of jacquards. The hook operation of mechanical jacquard invariably requires figured graph design as the first process and punching cards from the graph as the second process. In the textile industry, both of these processes were carried out manually.

In the recent past, electronic jacquards have been gradually replacing mechanical jacquards. Not only the organized weaving sector but also the decentralized power loom and handloom sectors of the textile industry have started gradually replacing mechanical jacquards with electronic jacquards. For operating the hooks of the electronic jacquard, the figured graph prepared in the computer with suitable weave marks is directly used without any card punching.

The Computer-Aided Figured Graph Designing (CAFGD), which was thus introduced for the electronic jacquard, has also become useful for the mechanical jacquard. After preparing the figured graph with suitable weave marks on the computer, the graph is printed. The cards are punched from the printed graph using hand-operated mechanical punching machines. Otherwise, without printing the graph, the figured graph prepared on the computer is directly used in the electronic card-punching machine to punch the cards. These punched cards are being used for operating the mechanical jacquards used in handlooms and power looms.

Thus, the CAFGD has become essential not only for the electronic jacquard weaving sector but also for the mechanical jacquard weaving sector. The manual figured jacquard graph designers working with mechanical jacquards have started shifting to computer-aided jacquard graph designing. Knowing the common principles of CAFGD for jacquard weaving is essential for working with any software available for this purpose. But, for learning the fundamentals, there is no book available as of date. Hence, the manual designers who intend to learn the knowledge of CAFGD mostly depend upon either the designers who are already in this field or the suppliers of computer-aided graph designing software.

Looking at this increasing knowledge thrust on CAFGD, the authors have come out to write a book on 'Computer-Aided Figured Graph Designing for Jacquard Weaving'.

For the last two decades, the authors have been working with all different kinds of software that are being used for computer-aided figured graph designing. Comparing all the software, the authors aim to give the systematic fundamental common procedure for doing computer-aided graph designing for jacquard weaving using any software.

This book aims to address the following areas:

- The scenario of jacquard graph designing before and after the invention of computer-aided jacquard graph designing

- Different terminologies of computers related to jacquard graph designing and hardware used
- The working principles of mechanical and electronic jacquards; harness building
- Algorithm to do figured graph designing manually and also using the computer
- How to create the figured design in different forms, arrangements, bases, and principles; also, how to calculate the graph count, graph size, and figure size on the fabric
- Scanning parameters, scanning the motif, colour reduction, colouring, and colour editing of the scanned motif
- Methods followed in altering the graph size, editing the graph, repeat mode drawing, repeat setting, and colouring
- Creating different weaves and saving them in the library, mapping the weaves with the edited image, and float-checking
- Developing figured graphs for different fabric varieties, taking the output of the graph, card punching, and taking digital output for electronic jacquards
- Advantages of computer-aided graph designing, different modules of graph designing software, and fabric simulation and draping

Since this book covers the fundamentals of CAFGD, which are common for all software, it will be useful for the designers to understand any software. This book will be a reference for manual designers who intend to become computer-aided designers, manual card-punching technicians who intend to become computer-aided card-punching technicians, mechanical jacquard manufacturers, and electronic jacquard manufacturers. This book covers the mathematical calculations to be followed in different stages of converting the given motif to a figured graph design with the required weave inputs. Hence, the designers, by going through this book, could be able to do figured graph designs for any variety of figured fabric from the given motif.

This book will be a textbook for the course on 'Computer-Aided Figured Graph Designing' for the Diploma, Post Diploma, and B.Tech. programs in Handloom, Textile, and Textile Designing Technology. This book will be a practical guidebook for the students to learn while practising and will serve as a reference guide to work with any software used in the jacquard weaving industry.

In the global textile industry, different jacquard manufacturers are producing mechanical jacquards and electronic jacquards under various brands. The jacquard fabric manufacturers use different software to prepare computer-aided figured graph designing for operating these machines. This book covers the common algorithms followed in preparing computer-aided figured graph designing, which makes global readers understand any global software easily after reading this book.

The contents of this book have been completely explained in simple grammatical English along with corresponding figurative illustrations to make the readers understand the principles of computer-aided figured graph designing clearly.

Acknowledgements

At the outset, we wish to acknowledge our thanks to the Indian Institute of Handloom Technology (IIHT), Salem, Tamil Nadu, Ministry of Textiles, Govt. of India, which made us knowledgeable, and the Department of Fashion Technology, KCG College of Technology, Karapakkam, Chennai, for encouraging and extending their support to write this book.

Firstly, we want to extend our heartfelt gratitude to our teachers Shri. M. K. Palanisamy, Shri. D. Jayaramaiah, Shri. G. Sukumaran Nair, Shri. C. Rajenthiran, and Late Shri. Rangachari of the IIHT, Salem, from whom we learned Woven Fabric Structures and Manual Figured Graph Designing for Mechanical Jacquards in our studies at IIHT, Salem during the tenure of Late Shri. L. Desi Gowda, former Principal.

We express our deepest appreciation from the bottom of the heart to Shri. G. Ramasamy, former Director, IIHT, Salem, and Shri. K.V. Mani, former Deputy Director, WSC, Bangalore, for driving us to learn, teach, and start writing the notes on 'Computer-Aided Figured Graph Designing (CAFGD)' two decades ago. Our thanks are due to Shri. G. Prabhakaran, Dean Academics, KCG College of Technology, Shri. V. Bhanu Rekha, Professor, and Shri. Mohamed Zakriya, Associate Professor, School of Fashion Technology, KCG College of Technology for their support and encouragement to write this book.

Also, we are thankful to our Mentors Shri. Kumar of Wovven Technologies and Shri. Tawassul Ansari of Texware Technology from whom we learned the fundamentals of CAFGD.

Furthermore, we wish to acknowledge our appreciation to Shri. N.S. Manokaran, Vocational Teacher, Textile Technology of Government Higher Secondary School, Onnupuram; the Master Weavers of Arani; and the Silk Saree Weavers of Durugam Village for making us know the figured graph designing for mechanical and electronic jacquards for weaving the high-end contemporary silk sarees.

Moreover, our gratefulness is to Shri. K. P. K. Jayaraja Shivam of Vasthra Design Creations, Arani, and Shri. Narendar Kolanu of Vastra CAD, Bangalore, for their support in exploring the different CAFGD software used in the industry.

Additionally, we are indebted to our family and friends for their unwavering support, encouragement, and understanding throughout the challenges and demands of this academic endeavour.

Finally, we wish to acknowledge our gratefulness for the guidance and support of the team of CRC Press and The Textile Institute in shaping and bringing this book into our hands.

1 Introduction to Computer-Aided Figured Graph Designing (CAFGD)

1.1 ELEMENTS OF WOVEN FABRICS

The fabrics produced in the textile industry are broadly classified into woven, knitted, and nonwoven fabrics. The woven fabrics are produced by weaving, the process of interlacing two series of threads; the knitted fabrics are produced by knitting, the process of inter-looping of yarns and inter-meshing of loops; and the nonwoven fabrics are produced by felting, the meshing of fibers together.

The simple coarser woven fabric and its elements are shown in Figure 1.1a. It consists of two series of yarns, longitudinal and horizontal, as represented in Figure 1.1b. The longitudinal yarn is called *the warp*, and the transverse yarn is called the *weft*. A single warp yarn is called the *end*. A single weft yarn is called the *pick*. While weaving the fabric, the pick is inserted by lifting and lowering the ends in a definite order. The order in which the ends are raised and lowered while inserting the pick is called the *interlacement*. End-up and end-down are the only possible interlacements. End-up interlacement is shown in Figure 1.1c. End-down interlacement is shown in Figure 1.1d. The fabric shown in Figure 1.1a is woven with a simple combination of these two interlacements. A repeat of these two interlacements on two ends and two picks, shown in Figure 1.1e, is called as *plain weave*. Figure 1.1f is the *Graph design* of the plain weave drawn on the point paper/graph paper. In this point paper, each vertical space between the lines denotes an end, and each horizontal space between the lines is a pick. The end-up interlacement of the weave is represented as a *mark*.

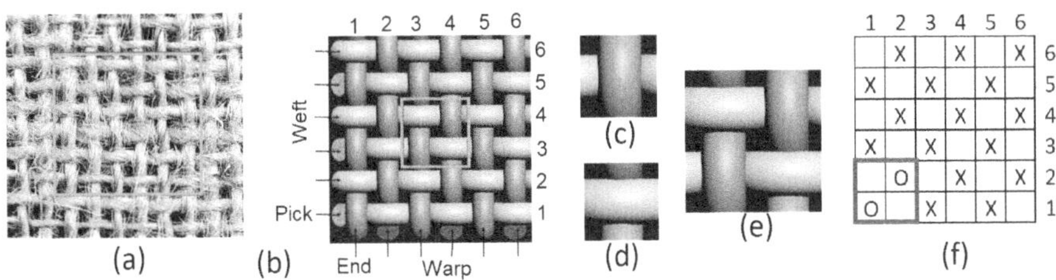

FIGURE 1.1 Simple woven fabric and its elements: (a) coarser woven fabric, (b) warp and weft, (c) end-up interlacement, (d) end-down interlacement, (e) interlacement of plain weave, and (f) graph design of plain weave.

DOI: 10.1201/9781003441205-1

The end-down interlacement is represented as *blank*. Hence, the graph design is the top view of the weave, showing the up-and-down interlacing of ends and picks using marks and blanks. The graph design is shown for six ends and six picks, with the repeat of the weave on two ends and two picks marked separately.

Interlacing each pick with warp threads to form the fabric is called the *weaving process*. The machine used for the weaving process is the *loom*. Shedding, picking, and beating are the primary motions of the weaving process. *Shedding* is the motion of separating the warp sheet into two layers, one above the other. The space formed between two layers is known as a *shed*. *Healds* or *harnesses* are used to make the shed of ends. Inserting or propelling the pick between the space formed by the shedding motion is called *picking*. A shuttle with the weft pirn is used for picking the weft. Beating the pick closer to the previous one (fell of cloth) using a reed fixed in the sley is called *beating*. Hence, in the loom, performing these three motions—shedding, picking, and beating—one after the other in cyclic order by keeping the warp sheet in a tight condition is called weaving. If the weaver uses his hands and legs to carry the primary motions, it is called the handloom or foot-treadle loom. When the rotation of the motor power is used to perform these motions, it is called the power loom. In shedding motion, to perform the shed (separating the warp sheet into two layers), the ends are controlled in groups through the healds or individually through the harnesses. The device or equipment used to operate the healds or harnesses is called a *shedding device*. The healds control the ends in groups to weave all-over and brick motif fabrics of simple or compound weaves. The harnesses control the ends individually to weave figured fabrics. A minimum of two healds are required to weave the all-over fabric of simple plain weave. A maximum of 16 healds can be conveniently used on a loom to weave other simple and compound weaves.

The sheds formed by the shedding device are of three types. They are bottom-close shed, centre-close shed, and open shed. All the shedding devices are set to produce any one of these sheds. In the bottom-close shed, when the loom is not in operation, the warp stays at the bottom line of the reed and on the sley race. While weaving, the top layer of the shed is formed by lifting the selective healds or harnesses from bottom to top and again lowering them from top to bottom after inserting the pick. In the centre-close shed, when the loom is not in operation, the warp stays at the centre line of the reed. While weaving, the top layer of the shed is formed by lifting the selective healds or harnesses from centre to top. Simultaneously, the bottom layer of the shed is formed by lowering the other healds or harnesses from centre to bottom. Both the layers come back to the centre of the reed after inserting the pick. In the open shed, when the loom is not in operation, one layer stays on the top and the other stays at the bottom. While weaving, selective healds or harnesses move from bottom to top. Simultaneously, selective healds or harnesses move from top to bottom.

1.2 CLASSIFICATION OF WEAVES

Woven fabric designers have already derived the number of weaves by combining the interlacements in various orders. The weaves thus derived are broadly classified into three categories: simple, compound, and pile, according to the number of warp and weft series used and the effect produced.

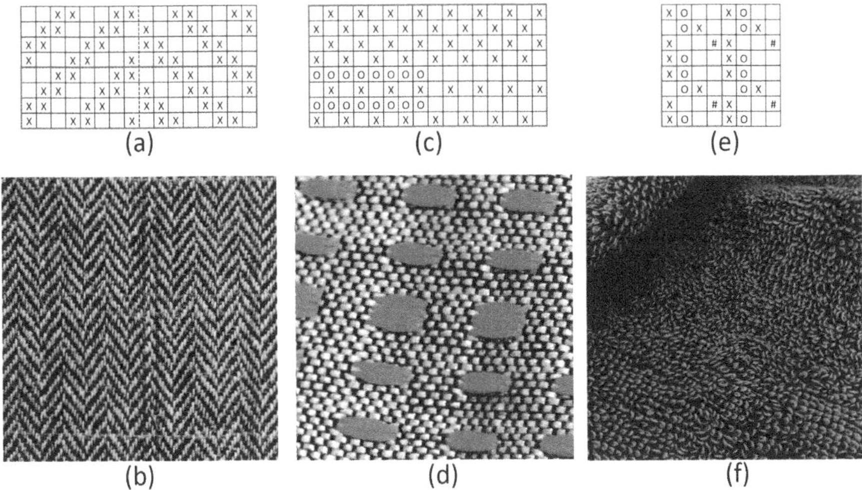

FIGURE 1.2   Classification of weaves: (a) simple weave, (b) simple weave fabric, (c) compound weave, (d) compound weave fabric, (e) pile weave, and (f) pile weave fabric.

1. Simple weaves: The weaves produced using a single warp and weft series are called simple weaves. Simple weaves include the three basic weaves: plain, twill, and sateen. Other weaves are derived and constructed based on these basic weaves. A few examples of simple weaves derived from the basic weaves are rib, mat, wavy twills, broken twills, herringbone twills, diamond, diaper, honeycomb, huck-a-back, and mock leno. For example, Figure 1.2a shows two repeats of a simple herringbone twill weave, and Figure 1.2b shows the herringbone fabric.

2. Compound weaves: Compound weaves are produced using more than one series of warp and weft arranged in layers. A few examples of compound weaves are bedford cord, welt, pique, backed cloth, extra warp, extra weft, double cloth, and patent satin. Figure 1.2c shows a part of the compound extra weft weave, and Figure 1.2d shows a furnishing fabric woven with this weave.

3. Pile weaves: Pile weaves produce loops of warp and weft yarns called *piles* that project from the foundation of the cloth woven in basic weaves. Terry, velvet, velveteen, and corduroy are a few examples. Figure 1.2e shows four repeats of a terry pile weave, and Figure 1.2f shows a towel fabric woven with this weave.

1.3   VARIETIES OF WOVEN FABRICS

In the weaving industry, three varieties of woven fabric structures are produced using simple, compound, and pile weaves. They are:

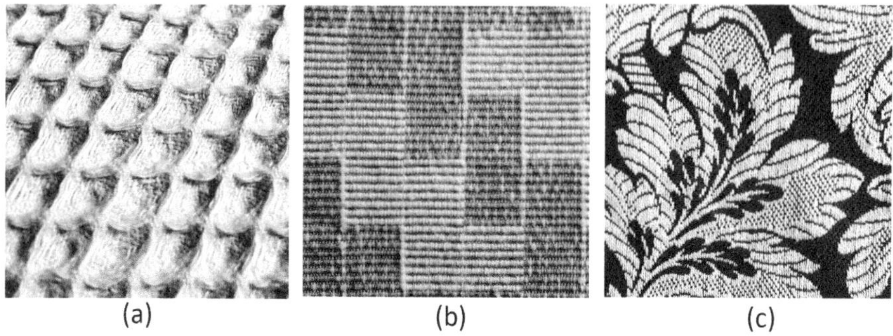

FIGURE 1.3   Varieties of woven fabrics: (a) all-over single-weave fabric, (b) geometric-brick motif fabric, and (c) figured motif fabric.

1. All-over single-weave fabric: This variety is woven using only one weave to produce a similar appearance all over the fabric surface. Figure 1.3a shows a fabric woven with only a honeycomb weave. A cellular structure is produced all over this fabric, similar to a honeycomb. This variety is woven using a multi-heald shedding device.
2. Geometric-brick motif fabric: This variety is produced by combining the two opposite weaves derived from one weave. In this fabric, each weave alternately repeats for a group of ends and picks for weaving simple figures with geometric-brick forms. Figure 1.3b shows a fabric woven with twilled brick form by combining the two derivatives of warp rib weave by changing the lifting of colouring ends arranged in 1:1 order. A multi-heald shedding device is used to weave this variety.
3. Figured motif fabric: This variety is woven by combining two or more weaves to produce distinct ground and figure parts of the curved motifs. For example, Figure 1.3c shows a white flower motif on a red ground surface. For weaving the figured fabrics, a jacquard shedding device is used, which operates the ends individually to form the interlacement as per the figure graph prepared.

1.4   HEALD-TREADLE AND HEALD-DOBBY SHEDDING

Primitive weavers improvised an arrangement to divide the warp ends into two layers, called *heald*. The most primitive heald consists of a series of twine loops, each passing around on alternate ends of the warp sheet and fastened to a wooden rod. The next alteration is of Indian origin and is 2000 years old. It is made by linking or clasping a second loop to the first loop. This simple contrivance increased the productivity of the loom. The healds were operated by the weavers' feet by using treadles. His hands were free to control other parts of the machine. The *Foot-Treadle Handloom* was the first technique for weaving fabrics. This technique prevails even today in many weaving centers.

In *heald shedding*, the number of healds required for weaving any weave is equal to the different number of interlacing ends present in that weave. Two healds and two treadles are used to weave plain fabric. Multi-healds and treadles are used to weave all-over fabrics and brick motif fabrics with simple, compound, and pile weaves. The treadles can conveniently operate up to 16 healds in the handloom and five healds in the power loom. A loom with four healds and treadles shedding is shown in Figure 1.4a. According to the lifting plan of the weave, a rope tie-up connection is arranged between healds and treadles. In the power loom, the healds are operated by the 'Tappets' through the treadles. Hence, it is named 'Heald-Tappet-Treadle Shedding'. From two to five healds are operated using two to five tappets along with roller reversing motion or spring reversing motion for weaving basic plain, twill, and sateen weaves.

A shedding device called *dobby* was invented to overcome the difficulty of operating more than five healds using treadles. The 'mechanical vertical hook dobby' shedding device was introduced in handloom to operate the healds. The 'mechanical horizontal lever dobby' shedding device was also introduced. In the handloom dobby, wooden horizontal levers are used to operate the harness, and in the power loom dobby, iron horizontal levers are used to operate the healds. In these dobbies, the hooks or levers are operated by lattices and pegs. Each hook or lever in the dobby operates a heald. In the handloom, the weavers operate the treadle connected to the lifting element of the hooks or levers. In the power loom, motor power does this operation. The hooks or levers to be lifted are selected through the pegs fixed on the lattices as per the weave to be produced. Each lattice represents one pick in the fabric.

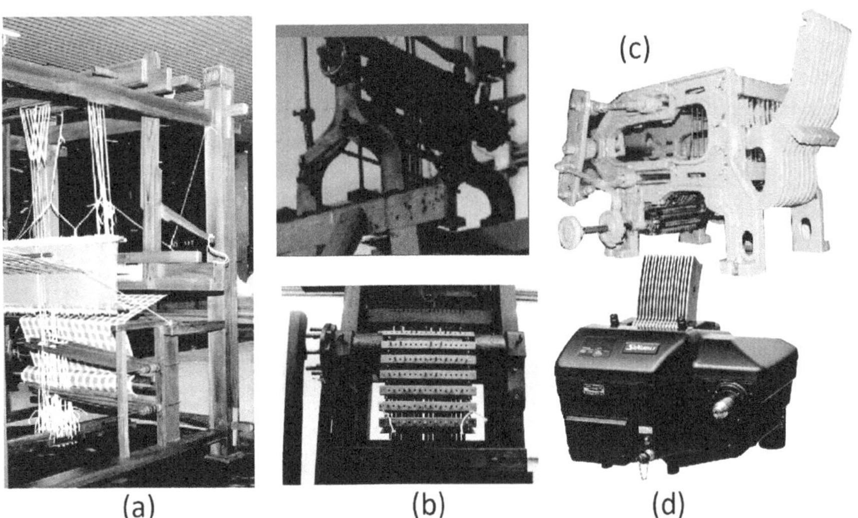

FIGURE 1.4 Heald-treadle and heald-dobby shedding: (a) heald-treadle shedding in the handloom, (b) vertical hook-heald dobby (mechanical) mounted on the top of the handloom with pegged lattices, (c) horizontal lever-heald dobby (mechanical) used in the power loom, and (d) horizontal lever-heald dobby (electronic) used in power loom.

The dobby requires a minimum number of lattices to weave all-over fabrics of simple weaves containing a minimum number of picks per repeat. However, huge lattices are required to produce brick motif fabrics in simple and compound weaves. Followed by the mechanical dobby, the electronic dobby was introduced. A set of solenoids and other electronic elements are used to lift the levers. The digital peg plan prepared in the dobby designing software activates these solenoids. A vertical hook dobby and the wooden lattices with pegs used to operate the hooks are shown in Figure 1.4b. A power loom dobby with iron horizontal levers is shown in Figure 1.4c. Figure 1.4d shows the electronic dobby used in the automatic power loom. Heald-dobby shedding is used for weaving all-over and brick motif fabrics with simple, compound, and pile weaves.

1.5 MECHANICAL AND ELECTRONIC JACQUARD SHEDDING

The ancient Jala/Adai, the first traditional shedding technique of figured weaving, is shown in Figure 1.5a. In this method, the ends are operated in figurative order by the series of thread connections, which are pre-interlaced as per the figured graph design. While weaving, the assistant weaver helps to form the shed; the weaver performs weaving. It was a time-consuming process. As time progressed, there was more and more demand for varied-figured materials. Mechanical geniuses had thoughts of devising a new shedding machine to perform the weaving operation faster to meet the increasing demand for the figured fabrics. The famous 'Jacquard loom' was the answer to these early-day difficulties.

'Joseph Marie Jacquard' received credit for inventing the machine that bears his name. The jacquard shedding device was a masterpiece of perfection in the textile world. The jacquard machine can weave practically any figure into the fabric. The jacquard shedding device is mounted on the top of the loom. The vertical hooks and the horizontal needles are the main parts of this machine. The jacquard hooks and the ends in the warp are connected with the fine twine called the harness. The figured graph is prepared, and cards are punched as per the marks and blanks on the graph. The punched card selects the hooks through the needles. The lifting mechanism lifts the selected hooks, thereby

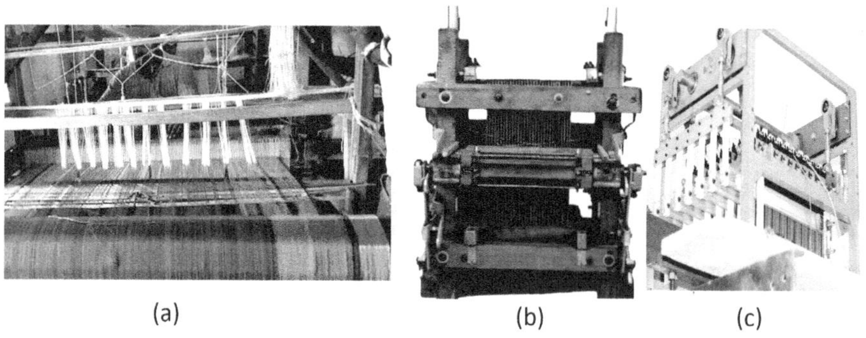

(a)   (b)   (c)

FIGURE 1.5 Shedding devices used for figured fabric weaving: (a) traditional Jala/Adai shedding method, (b) wooden mechanical jacquard, and (c) electronic jacquard.

lifting the ends as per the figuring interlacement prepared on the graph. As all the parts of this machine work on a mechanical principle, it is called 'mechanical jacquard'. The wooden mechanical jacquards used in handlooms for weaving traditional silk sarees are shown in Figure 1.5b.

Bonas Textile Machinery NV launched the first successful 'Electronic Jacquard' at ITMA Milan in 1983. In the recent past, the electronic jacquard has been introduced gradually to replace the above-mentioned mechanical jacquard. The electromagnetic (solenoids) principle is used instead of needles and cards to operate the hooks. The selection of solenoids is directly done by the digital figured graph design prepared in the computer. Hence, preparing the figured graph manually on the graph paper and punching the cards from the graph are eliminated. The hook-lifting mechanism works with the double-lift principle, which is operated by a servo motor. Figure 1.5c shows the electronic jacquard.

Table 1.1 compares the different shedding devices used to operate the healds and harnesses for weaving different fabric varieties along with the other parts and functions.

1.6　MANUAL FIGURED GRAPH DESIGNING

The process of mechanical jacquard weaving starts with preparing a figured graph manually. In the manual graph preparation, an enlarged outline of the figure is drawn in pencil on point paper. The figure parts are separated from the ground by converting the figure outline into a stepped line by following the grid lines of the point paper. Necessary weaves are inserted both on the figure and the ground. Figure 1.6a shows a part of the figured graph prepared manually on 40 ends × 32 picks.

Card punching is the next process after completing the figured graph. The weave marks in each pick of the graph are punched as holes in multiple rows on each card. Initially, the perforated *punching plate*, as shown in Figure 1.6b, was used for punching the holes in the card by hand. The punching is carried out by fixing the card between the two punching plates. The punching rod is moved over the card to punch hole after hole as per the marks in the figured graph.

Later, a punching machine was developed under the name of *piano card-punching/ cutting machine*. The punching box is an important part of this machine. The punching pins to punch cards and keys to operate these pins are arranged in this box. By pressing the pedal, the card moves back below this box. At the same time, the punching pin box moves up and down over the card. The operator presses the keys with his fingers as per the marks on the figured graph design. By doing so, the pins punch the holes on the card as per the marks on the figured graph design. Figure 1.6c shows the piano card-punching machine. As shown in Figure 1.6d, the punched cards are laced together into a continuous sequence to make a chain. The chain is mounted on the jacquard machine as shown in Figure 1.6e. The holes on the card operate the hooks through the needles. The hooks lift the ends in figuring order independently as per the graph design prepared, by which elaborate details and flowing lines are produced in a texture.

TABLE 1.1

Comprehensive Details of Different Shedding Devices Used to Operate Healds and Harnesses

Shedding Device/Parts and Function	Multi Treadle	Horizontal Lever Dobby	Vertical Hook Dobby	Electronic Dobby	Mechanical Hook Jacquard	Electronic Jacquard
Ends carrying part	Healds	Healds	Healds	Healds	Harnesses	Harnesses
Healds lifting part	Treadles	Horizontal levers	Vertical hooks	Horizontal levers	Vertical hooks	Vertical hooks
Selecting part of the lifting part	Tie-up	Pegged lattices	Pegged lattices	Electromagnet-solenoids	Needles and punched cards	Electromagnet-solenoids
Designing for the selection part	Lifting plan of the weave	Lifting plan of the weave	Lifting plan of the weave	Digital lifting plan in computer	Punched cards from figured graph	Digital figured graph in computer
Operating part	Multi treadles	Single treadle	Single treadle	Pedal switch /motor	Single treadle	Pedal switch/motor
Reversing motion used	Tumbler lever	Spring	Spring	Spring	Lingo	Lingo/Spring
Shed formed	Centre close	Bottom close	Bottom close	Bottom close	Bottom close	Open
Used in	Handloom	Handloom and power loom	Handloom	Handloom and power loom	Handloom and power loom	Handloom and power loom
Fabric variety produced	All-over and brick motif	All-over and brick motif	All-over and brick motif	All-over and brick motif	Figured motif	Figured motif

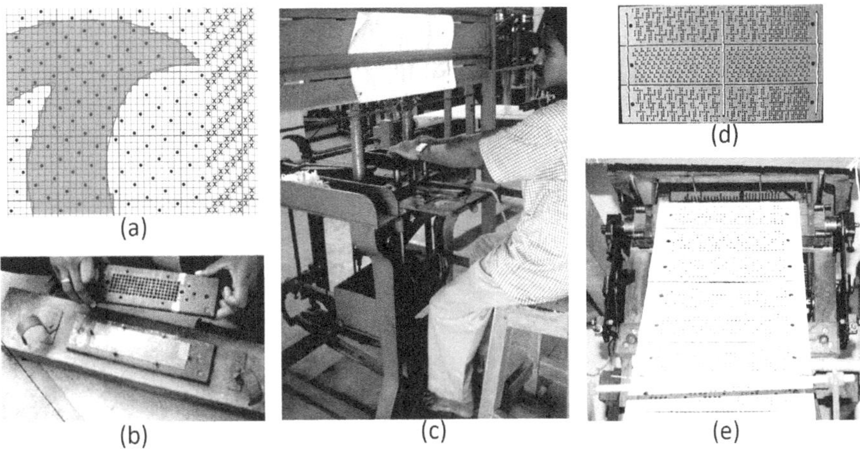

FIGURE 1.6   Manual figured graph designing, card punching, and lacing: (a) a part of a manual figured graph on 40 ends × 32 picks, (b) manual card-punching plate, (c) manual-mechanical piano card-punching machine, (d) punched cards laced in chain form, and (e) punched chain of cards mounted in the jacquard.

1.7   COMPUTER-AIDED FIGURED GRAPH DESIGNING (CAFGD)

In the textile industry, initially, the technologists developed the 'computer-aided figured graph designing (CAFGD)' software exclusively to do the digital figured graph design with marks and blanks for operating the electronic jacquard to produce elaborate figured fabrics. Figure 1.7a shows the computer window with the part of a figured graph on 32 ends × 32 picks. The digital design is loaded into the control box of the electronic jacquard. The black marks and blanks generate the signals and operate the magnets in the modules through the electronic control system and boards. The electromagnets carry out the selection or rejection of the hooks corresponding to the marks and blanks of the figured graph. The digital graph thus prepared using

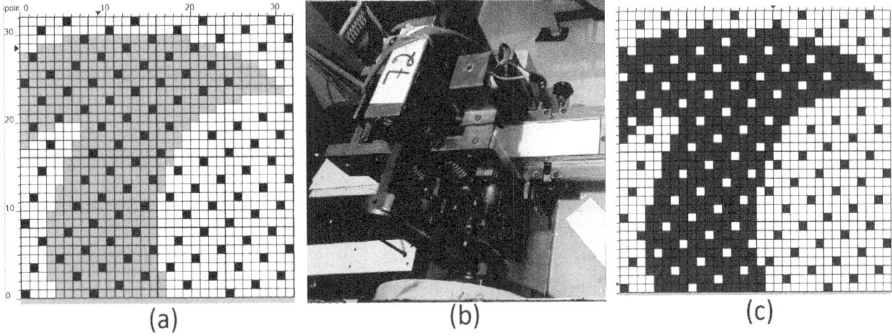

FIGURE 1.7   Computer-aided-figured graph designing and card punching: (a) computer window with the part of a figured graph on 32 ends × 32 picks, (b) an electronic card-punching machine, and (c) a printed graph on paper from a digital graph on a computer.

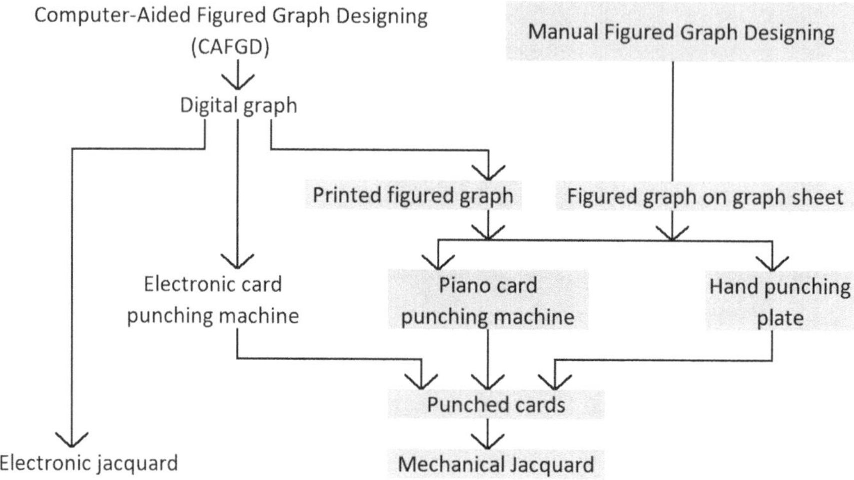

FIGURE 1.8   Flowchart showing the versatility of CAFGD over the manual figured graph designing.

the computer directly operates the electronic jacquard, thereby eliminating the graph paper and cards punched from the graph paper used for operating the mechanical jacquard.

Later, the designers started using the CAFGD for the mechanical jacquard too. The engineers developed the electronic card-punching machine, as shown in Figure 1.7b. This machine works on the principle of the piano card-punching machine to punch the cards. The electronic control system loaded with the digital figured graph operates the punches in this machine as per the marks and blanks of the digital figured graph. Furthermore, the digital figured graph is printed using a plotter or printer in a single colour, as shown in Figure 1.7c. Cards are punched from this printed graph paper using either the mechanical piano punching machine or a hand punching plate. These punched cards are used for operating the mechanical jacquard used in the handloom and power loom. Thus, the CAFGD has become versatile for both electronic and mechanical jacquards, as shown in the flow chart given in Figure 1.8. Therefore, learning CAFGD has become essential not only for the designers working with electronic jacquards but also for the designers working with mechanical jacquards.

BIBLIOGRAPHY

1. Grosicki, Z. J. (2004). Elements of woven design. In *Watson's Textile Design and Colour* (pp. 1–7). Woodhead Publishing Limited.
2. Grosicki, Z. J. (2004). Elements of jacquard shedding. In *Watson's Textile Design and Colour* (pp. 178–181). Woodhead Publishing Limited.
3. Mitra, A. (2014). CAD/CAM solutions for textile industry. *International Journal of Current Research and Academic Review.* 2(6), 41–50.
4. Panneerselvam, R. G., Yuvaraj, D., and Bhanu Rekha, V. (2020). Use of indigenous electronic jacquard in handloom for weaving fashionable silk sarees. *Dogo Rangsang Research Journal.* 10(7), 84–90.

2 Computer-Aided Figured Graph Designing— Terminologies

2.1 PIXEL, RASTER IMAGE, AND VECTOR IMAGE

The figured graph designers need to know about the different terminologies of the software and hardware commonly used in the graph designing field. It is not an exaggeration to say that the *jacquard machine* is the father of the computer machine. The computer screen can be compared to the point paper used for the figured graph designing. The idea of storing the data in the *storage devices* in *zeros* and *ones* came from the *holes* and *blanks* of the *punched cards* used in the jacquard machine to produce figured fabrics. The *electronic circuit* in the computer is the latest form of harness constructed (*harness building*) between the hooks of the jacquard and warp ends. Therefore, it is easy for graph designers to understand the terminologies of the software and hardware.

The *monitor screen* of the computer (the output device) is nothing but point paper that is used for jacquard designing, as shown in Figure 2.1a. Like point paper, the screen is divided into vertical and horizontal spaces. The rectangular/square area where the vertical and horizontal spaces intersect is called a pixel (an abbreviation of picture cell). Figure 2.1b shows the pixels of the screen. Each pixel on the screen is a unit. It is only possible to change the colour of each pixel in full, not in fractions. In Figure 2.1c, one pixel is filled with one colour. The *pixels per inch* refer to the resolution of the screen. The image made up of a collection of tiny squares/pixels/dots is called a raster image. It is also called a grid image, pixel image, dot image, bitmap image, or digital image. Figure 2.1d shows the raster image of a circle formed with colour pixels in the graph. When this raster image is zoomed in using the 'zoom in' option, individual pixels can be seen. The digital camera automatically shoots and

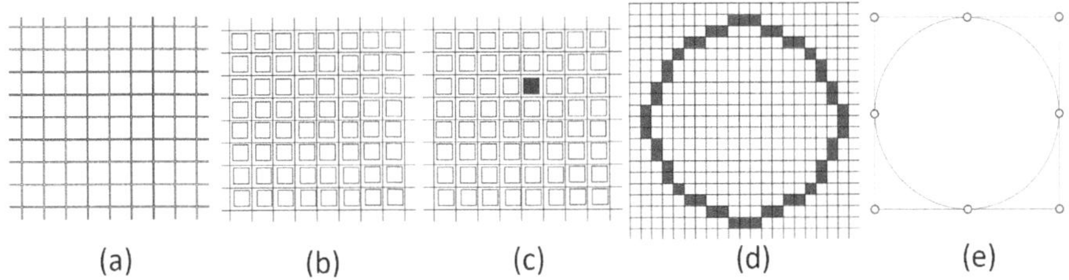

(a)   (b)   (c)   (d)   (e)

FIGURE 2.1 Pixel, raster image, and vector image: (a) point paper, (b) pixel of the screen, (c) one pixel filled with one colour, (d) raster/pixel image of the circle, and (e) vector image of the circle.

DOI: 10.1201/9781003441205-2

saves photos as raster images. The scanner also scans and saves the images on paper as raster images.

The image drawn on paper using lines or curves is called a vector image. On the computer, vector files use mathematical equations for lines and curves with fixed points on a grid to produce an image. There are no pixels in the vector image. The mathematical formulas of vector files capture shape, border, and fill colour to build an image. It is possible to recalibrate the mathematical formulas to any size. Hence, vector images can be scaled up or scaled down without impacting their quality. With vector image files, resolution is not an issue. Vector images can be resized, rescaled, and reshaped without losing the image quality. Vector files are popular for images that need to appear in a wide variety of sizes, like a logo that needs to fit on both a business card and a billboard. Figure 2.1e shows the vector image of the circle. It is drawn on the blank document using a circle drawing tool, keeping the diameter at 6 cm.

All the computer-aided figured graph designing (CAFGD) software used in the textile industry works the image on a raster base in graphical format. All the photo editing software, like MS Paint, Paint.net, and Paintshop Pro, works with figures only in pixel format. Corel Draw and Adobe Photoshop have the facility to do both in vector and raster formats. The Computer-Aided Designing (CAD) software used in mechanical and civil engineering designs the shapes in vector format.

2.2 PIXEL NUMBERING

On the graph sheet/point paper, each vertical space represents an end. Each horizontal space represents a pick. Graphically, each vertical space (end) crosses the x-axis. The horizontal space (pick) crosses the y-axis. Hence, in computer graph designing, 'x' denotes end and 'y' denotes pick. Each pixel—the intersecting point of the end and pick—is represented by the 'x' and 'y' coordinates (x, y). The horizontal and vertical pixels are given numbers serially to have the coordinates for the pixel. Figure 2.2a shows the pixel numbering followed in the photo editing and drawing software. The x-ends are numbered 0, 1, 2, 3...from left to right. It is denoted as x0,

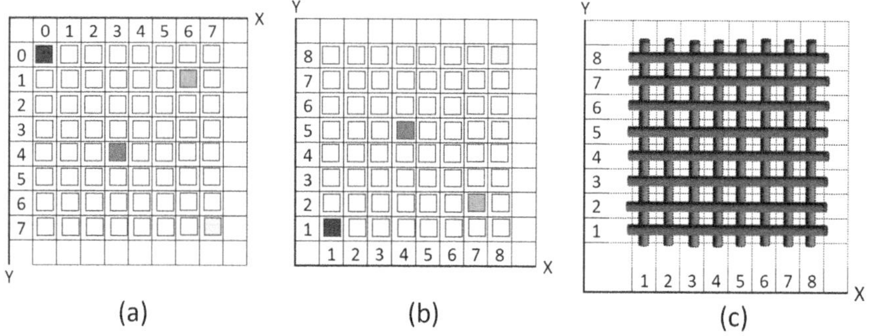

FIGURE 2.2 Pixel numbering: (a) pixel numbering in photo editing/drawing software, (b) pixel numbering in CAFGD software, and (c) imagination of eight ends and eight picks.

x1, x2, x3…xn. The y-picks are numbered 0, 1, 2, 3…from top to bottom. It is denoted as y0, y1, y2, y3…yn. The pixel number is denoted by (xn, yn). In Figure 2.2a, the coordinate of the red-colour pixel is (x0, y0) or just (0, 0). It represents the intersecting point of the first end and first pick (1, 1). The coordinate of the green-colour pixel is (6, 1). It is the intersecting point of the seventh end and second pick (7, 2). Thus, in photo editing software, the end and pick numbers are calculated by adding 1 to the pixel coordinate. Accordingly, the coordinates of the blue-coloured pixel (3, 4) show the intersecting point of the fourth end and fifth pick (4, 5). The coordinate of the yellow-coloured pixel (7, 7) denotes the intersecting point of the eighth end and eighth pick (8, 8).

Figure 2.2b shows the pixel numbering followed in the figured graph designing software. The x-ends are numbered 1, 2, 3…from left to right. The y-picks are numbered 1, 2, 3…from bottom to top. Hence, in Figure 2.2b, the coordinate of the red-colour pixel is (1, 1). It directly represents the intersecting point of the first end and the first pick (1, 1). The coordinate of the green-colour pixel is (7, 2). It represents the intersecting point of the seventh end and second pick (7, 2). Thus, in the figured graph designing software, the pixel coordinate directly indicates the end and pick numbering. Accordingly, the coordinates of the blue-colour pixel (4, 5) show the intersecting point of the fourth end and fifth pick (4, 5). The coordinate of the yellow-colour pixel (8, 8) denotes the intersection of the eighth end and eighth pick (8, 8). Figure 2.2c shows the eight ends and eight picks that the designer has to imagine corresponding to the eight rows and eight columns shown in Figure 2.2b.

2.3   RESOLUTION OF THE MONITOR SCREEN

On any monitor screen, the number of pixels in its width (in one row) and the number of pixels in its length (in one column) denote the screen display resolution. For example, the screen resolution is given as 1792 × 1344, as shown in Figure 2.3. It means that the screen contains 1792 pixels in its width (in each row) and 1344 pixels in its length (in each column). The total pixels on the screen are 24,08,448 (1792 × 1344). Then, the resolution in pixels per square inch is calculated by noting the screen width and length in inches. If the screen size is 12.8″ in width by 9.6″ in length, the 'pixels per inch' is calculated as below:

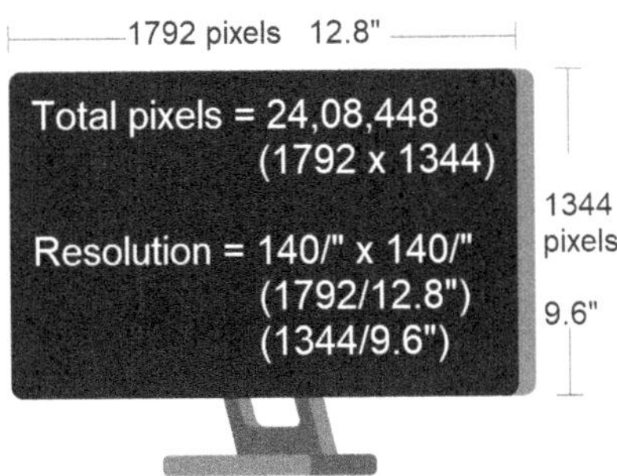

FIGURE 2.3   Calculating the resolution of the monitor screen.

$$\text{Pixels per inch in width} = \frac{\text{Total pixels in width}}{\text{Screen width in inches}} = \frac{1792}{12.8} = 140$$

$$\text{Pixels per inch in length} = \frac{\text{Total pixels in length}}{\text{Screen length in inches}} = \frac{1344}{9.6} = 140$$

In one square inch of the screen, there are 140 pixels by 140 pixels, which is denoted as 140 PPI (pixels per inch) resolution. It is possible to decrease the resolution of the screen from 1792 × 1344 to 1024 × 768. Therefore, for example when the screen display is changed to:

1536 × 1152 Resolution, the PPI becomes 120 [(1536 pixels/12.8″) × (1152 pixels/9.6″)],

1280 × 960 Resolution, the PPI becomes 100 [(1280 pixels/12.8″) × (960 pixels/9.6″)], and

1024 × 768 Resolution, the PPI becomes 80 [(1024 pixels/12.8″) × (768 pixels/9.6″)]

The higher the resolution of the screen, the smaller the icon/picture appears on the screen, and vice-versa. An icon, which is drawn by the software programmers on 70 × 70 pixels size, looks 0.5″ × 0.5″ size on the screen if the resolution is set to 140 PPI. If the screen resolution is changed from 140 to 100 PPI, the same icon looks 0.7″ × 0.7″ size on the same screen.

2.4 COLOUR THEORY AND PIXEL COLOUR

At its core, colour is our perception. When we see anything, the data sent from our eyes to our brain interprets it as a particular colour. Objects reflect light in different combinations of wavelengths (radio, gamma rays, and more). Our brains pick up on these wavelength combinations, interpreting them as a phenomenon known as 'Colour'. Colour theory is the art and science of using colour. It explains how humans perceive colour both physically and psychologically and how colours mix, match, and contrast with one another.

Light colour theory, pigment colour theory, and printer colour theory are the three different colour theories replicated in computers, printing, and the colouring of textiles, respectively. Table 2.1 highlights the features of these colour theories for comparative understanding.

Hence, it is worth noting that the light colour theory and printer colour theory are opposed to each other. When the colour image on the screen is taken as a printout, the RGB values of different colours that appear on the screen are first converted into corresponding CMYK values (Ex. 73 R + 73 G + 85 B = 12% C + 12% M + 0% Y + 15% K). In the printer, CMYK colours are mixed to print the same colour as it appears on the screen. Of course, the perfect match between the colours on the screen and the colours printed depends on the CMYK colours available in the cartridge.

TABLE 2.1

Comparison between Light, Printer, and Pigment Colour Theories

Light Colour Theory	Printer Colour Theory	Pigment Colour Theory
The colours that appear on the screen by a cathode ray tube (CRT) are based on the light theory.	The colour inks used in the printer are based on the principle of printer colour theory.	The dye pigments used by the dyers in colouring the textile are based on the pigment theory.
Red (R), Green (G), and Blue (B) are the three primary colours. Hence, it is RGB colour theory.	Cyan (C), Magenta (M), and Yellow (Y) are the primary colours. Black (K) colour is also used along with primary colours. Hence, it is CMYK colour theory.	Red (R), Yellow (Y), and Blue (B) are the three primary colours. Hence, it is RYB colour theory.
The secondary colours are: R + G = Yellow (Y) G + B = Cyan (C) B + R = Magenta (M)	The secondary colours are: C + M = Blue (B) M + Y = Red (R) Y + C = Green (G)	The secondary colours are: R + Y = Orange (O) Y + B = Green (G) B + R = Purple (P)
These colours are shown in Figure 2.4a	These colours are shown in Figure 2.4b.	These colours are shown in Figure 2.4c
The primary colours of the light theory are the secondary colours of the printer theory.	The primary colours of the printer theory are the secondary colours of the light theory.	–
The depth values of each colour are represented by a number from 0 to 255 (256 variations).	The depth values of each colour are represented by a number from 1 to 100 (100 variations).	The depth values of each colour are represented by the percentage of colour based on the weight of the material used for colouring.
The 0 value of red means Black.	The 0 value of cyan means White.	2% red means 2 grams of red Pigment is needed to colour 100 grams of material.
The 255th value of red means Full Red.	The 100th cyan means full Cyan.	
When no light colour is present, it is black.	When no printer ink is present, it is white.	When no pigment is present, it is white.
0 R + 0 G + 0 B = Black	0 C + 0 M + 0 Y = White	0 R + 0 Y + 0 B = White
When all three primary colours are present in their highest value, it results in white.	When all three primary colours are combined, it results in dark brown. A small quantity of black turns the dark brown colour into black.	When all three primary colours are combined, it results in dark brown. A small quantity of black turns the dark brown into colour black.
255th R + 255th G + 255th B = White	100 C + 100 M + 100 Y + Little Black = Black	4% R + 4% Y + 4% B + Little Black = Black
It is the additive colour theory (Black to White)	It is the subtractive colour theory (White to Black)	It is the subtractive colour theory (White to Black)

(*Continued*)

TABLE 2.1 (*Continued*)

Comparison between Light, Printer, and Pigment Colour Theories

Light Colour Theory	Printer Colour Theory	Pigment Colour Theory
Grey-tone colours are obtained by combining the similar values of three primary colours.	Grey-tone colours are obtained by mixing the similar values of three primary colours.	Grey-tone colours are obtained by mixing the similar values of the three primary colours.
54th R + 54th G + 54th B = Dark Grey	25th C + 25th M + 25th Y = Light Grey	By mixing 0.1% of all three primary colours, a light grey colour is obtained.
128th R + 128th G + 128th B = Medium Grey	50th C + 50th M + 50th Y = Medium Grey	By adding 1% of all three primary colours, a dark grey colour is obtained.
180th R + 180th G + 180th B = Light Grey	75th C + 75th M + 75th Y = Dark Grey	
Different colours are obtained by combining different values of three primary colours.	Different colours are obtained by adding different values of three primary colours.	Different colours are obtained by adding different percentages of three primary colours.
200th R + 100th G + 50th B = Greenish Red	80th C + 40th M + 20th Y = Greenish Blue	4% R + 1% Y + 0.25% B = Greenish Red
50th R + 200th G + 100th B = Bluish Green	20th C + 80th M + 40th Y = Bluish Red	0.25% R + 4% Y + 1% B = Purplish Yellow
100th R + 50th G + 200th B = Reddish Blue	40th C + 20th M + 80th Y = Reddish Green	1% R + 0.25% Y + 4% B = Orangish Blue
These colours are shown in Figure 2.4d.	These colours are shown in Figure 2.4e.	
The total number of colours possible is $256 \times 256 \times 256 = 16.7$ million colours, which are called *True colours*.	The total number of colours possible is $100 \times 100 \times 100 = $ one million colours	Infinite number of colours are possible
Out of 16.7 million colours, when a few colours are selected (say 256), it is called *Indexed colours*.		

The number of colours in which the pixels appear on the screen depends on the number of 'Bits' working for each pixel. One bit represents one unit of space occupied in the storage device. The relation between the number of colours and bits is given by: The number of colours is 2^n, where n is 'Bits Per Pixel'. When the designer set the computer to 1 bit per pixel, the number of colours in each pixel is $2^n = 2^1 = 2$. This means it is possible to have each pixel only in two colours: black and white. The pixel could be either white or black. When the designer- set the computer to 8 bits per pixel, the number of colours is $2^n = 2^8 = 256$. This means, it is possible to have 256 colours in the computer palette, and it is possible to change each pixel into any one of these 256 colours. Therefore, on the computer,

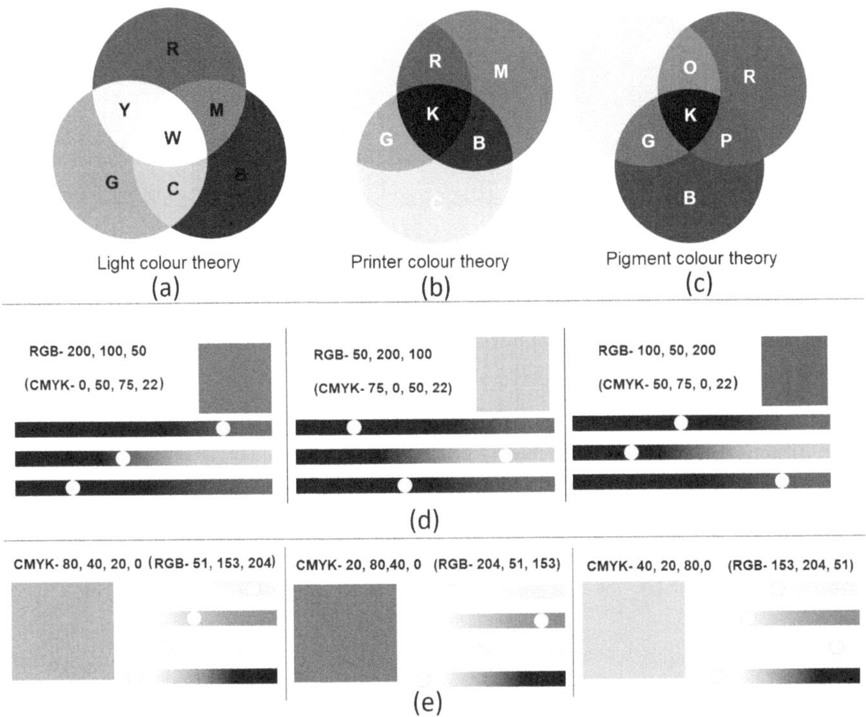

FIGURE 2.4 Light, printer, and pigment colour theories: (a) light colour theory, (b) printer colour theory, (c) pigment colour theory, (d) different RGB mixing of light colour theory, and (e) different CMYK mixing of printer colour theory.

1 bit /pixel – image = 2^1 = 2 colours image (monochrome bit map)
2 bits/pixel – image = 2^2 = 4 colours image
4 bits/pixel – image = 2^4 = 16 colours image
8 bits/pixel – image = 2^8 = 256 colours image (*Indexed Colour*)
16 bits/pixel – image = 2^{16} = $2^8 \times 2^8$ = 256 × 256 = 65,536 colours image
24 bits/pixel – image = 2^{24} = $2^8 \times 2^8 \times 2^8$ = 256 × 256 × 256 = 16. 7 million
    colours image (*True Colour*)

2.5 GRAPH SIZE ON THE SCREEN

In the CAFGD software and editing software, opening the new file means opening the new graph sheet. To open the new graph, the width pixels and the length pixels are given as input in the new file dialogue box. The width pixels are the ends of the graph, and the length pixels are the picks of the graph. The size of the new file that gets opened on the screen depends upon the resolution (PPI) of the screen. For example, to open the graph with 240 ends and 120 picks, the input given is 'width pixels 240' and 'length pixels 120' as shown in Figure 2.5a. Considering that the

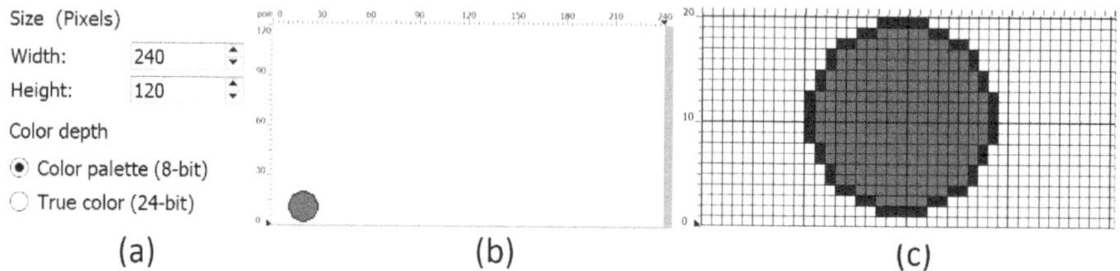

FIGURE 2.5 Graph size on the screen: (a) opening a new graph of 240 width pixels (ends) × 120 height (length) pixels (picks), (b) file opened as a white paper of 2″ width ×1″ length size on the 120 PPI screen, and a circle of 19 ends ×19 picks is drawn at the bottom left corner, and (c) the white paper made into a graph with 1000% zoom, and the circle drawn at the bottom left corner is seen in pixels of 19 ends ×19 picks.

resolution of the screen is 120, the size of the new file opened on the screen is calculated as given below:

$$\text{Width of the new file on the screen in inches} = \frac{\text{Total width pixels}}{\text{PPI}} = \frac{240}{120} = 2''$$

$$\text{Length of the new file on the screen in inches} = \frac{\text{Total length pixels}}{\text{PPI}} = \frac{120}{120} = 1''$$

the width × length of the new file opened is 2″ × 1″, as shown in Figure 2.5b. This 2″ × 1″ size is the motif size on the screen. By zooming the file to 1000% (10 times) using the zoom tool, these 240 × 120 pixels get enlarged to a 24″ × 12″ size with 10 PPI.

When the grid/graph lines are put on with the 'grid on' tool, this pixel area becomes a graph sheet of 240 ends × 120 picks. It is important to note that before zooming in, the 240 pixels × 120 pixels in 2″ × 1″ size is the paper/motif size. After zooming in, the very same motif becomes the graph of that motif on 240 ends × 120 picks size. In Figure 2.5b, a small circle drawn at the bottom of the motif on 19 ends × 19 picks size is the raster or pixel image. Figure 2.5c shows 40 ends × 20 picks of this graph (the bottom left part). The pixels of the same image are seen ten times enlarged between the graph lines, as shown in Figure 2.5c.

2.6   MEMORY SIZE OF THE GRAPH IN THE COMPUTER

Consider a jacquard graph design has to be prepared on 600 ends × 400 picks. It is drawn by opening a new file and taking 600-width pixels × 400-length pixels. If the new file is opened, keeping the colour option at 24 bits per pixel, the colour palette is opened with true colours (16.7 million colours). Similarly, the colour palette will be in indexed colour (256 colours) if 8 bits per pixel are taken. It is also possible to change the colour option after opening the file. Again, while saving the file, the required colour option can be selected.

Consider that the graph is prepared using only the black colour over the white ground and requires to save the same with these two colours. Then, the file is saved with the monochrome bit map (1 bit per pixel = 2 colours) colour option. If so, the memory size occupied by this graph in the storage device is calculated as given below:

Total number of pixels = width in pixels × length in pixels = 600 × 400 = 240,000
Number of bits required = Total pixels × 1 bit/pixel = 240,000 × 1 = 240,000
Memory space required = 240,000 bits = 30,000 bytes = 29.29 kilobytes (KB)
(8 bit = 1 byte; 1024 bytes = 1 kilobyte; 1024 kilobytes = 1 megabyte – MB)

Therefore,

$$\text{Memory size occupied in KB} = \frac{\text{Width pixels} \times \text{length pixels} \times \text{bits per pixel}}{8 \times 1024}$$

$$= \frac{600 \times 400 \times 1}{8 \times 1024} = 29.29 \text{ KB}$$

Thus, a 600 × 400 graph design in black and white colours occupies 30 KB of memory. When the same graph is prepared by keeping 256 indexed colours in the palette (8 bits per pixel) and saved by taking the colour option with 256 indexed colours (8 bits per pixel), the memory size occupied for this design in the storage device is:

$$\text{Memory size occupied in KB} = \frac{600 \times 400 \times 8}{8 \times 1024} = 234.4 \text{ KB}$$

When the same 600 × 400 design is prepared and saved by keeping 16.7 million true colours (24 bits per pixel) in the palette, the memory size occupied for this design in the storage device is:

$$\text{Memory size occupied in KB} = \frac{600 \times 400 \times 24}{8 \times 1024} = 703 \text{ KB}$$

It is also possible to save the file in indexed colour, which is opened with true colours, and vice-versa. In this case, the computer gives an alert that the colour option is changing.

BIBLIOGRAPHY

1. Grosicki, Z. J. (2004). Elements of colour. In *Watson's Textile Design and Colour* (pp. 130–148). Woodhead Publishing Limited.
2. Enriquez, R. F. (2020). Basics of computer graphics. In *Basics of Computer Graphics and An Introduction to Graphic Design* (pp. 1–10). Creative Hands Publishing.
3. Thareja, R. (2019). *Fundamentals of Computers*. Oxford University Press.
4. Kuehni, R. G. (2015). *Color – An Introduction to Practice and Principles*. John Wiley & Sons. Inc. Publication.

3 Working Principle of Mechanical and Electronic Jacquards

3.1 MECHANICAL JACQUARD—WORKING PRINCIPLE

The term 'jacquard' is not exact or partial to any particular loom but slightly refers to the additional managing instrument that automates the patterning of weaving. Jacquard is a Design box or Design head. Joseph Marie Jacquard was the inventor of jacquards in the 18th century, hence the name jacquard. If a fabric contains a vast number of ends bound for different interlacements with the weft, the method of guiding the warp by heald frames is too cumbersome and inefficient. In such cases, it becomes necessary to use the jacquard machine to raise each end independently using a hook, needle, card, and harness. As the parts of the jacquard machine are mechanically operated, the jacquard is called a 'mechanical jacquard'.

Figure 3.1 shows the working principle of mechanical jacquard. Figure 3.1a shows the basic setup of different parts of the jacquard to lift the end. As shown in the figure, each end of the warp sheet is drawn through the mail in the harness (ha). Each jacquard hook (h) is 16″ to 17″ long, with a small bend at the top and a long bend at the bottom. A neck cord is fastened at the lower bend. The neck cord is connected to a bunch of harness cords (co). The number of harness cords per bunch is equal to the number of repeats in the full width of the warp. The harness cords are passed through the holes in the comber board (cb). Each harness cord is connected to the upper coupling of the harness (ha). At the end of the upper coupling of the harness, there is harness mail through which the end is drawn. The bottom of the harness mail is tied with the lower coupling. At the end of the lower coupling of the harness, a thin rod weighing 5–15 g called 'lingo' is tied to bring down the end when the hook comes down. The connection between the hook and the harness is made in such a way that the ends touch the bottom of the reed to enable it to form a 'bottom close shed'. The purpose of drawing the harness cords through the holes of the comber board is to keep the number of harness cords in a unit space equal to the number of ends per unit space and the width of the harness in the comber board is equal to the width of warp in the reed.

Raising the hooks in the jacquard machine raises the harness cords, and the latter raises every end throughout the fabric width for interlacing with the pick. A horizontal needle (n) controls the hook in the jacquard. The needle on one end faces the punched card and the other end is pressed by the spring (s). For lifting the hook, a knife bar (k) is provided just below the hook's top bend.

DOI: 10.1201/9781003441205-3

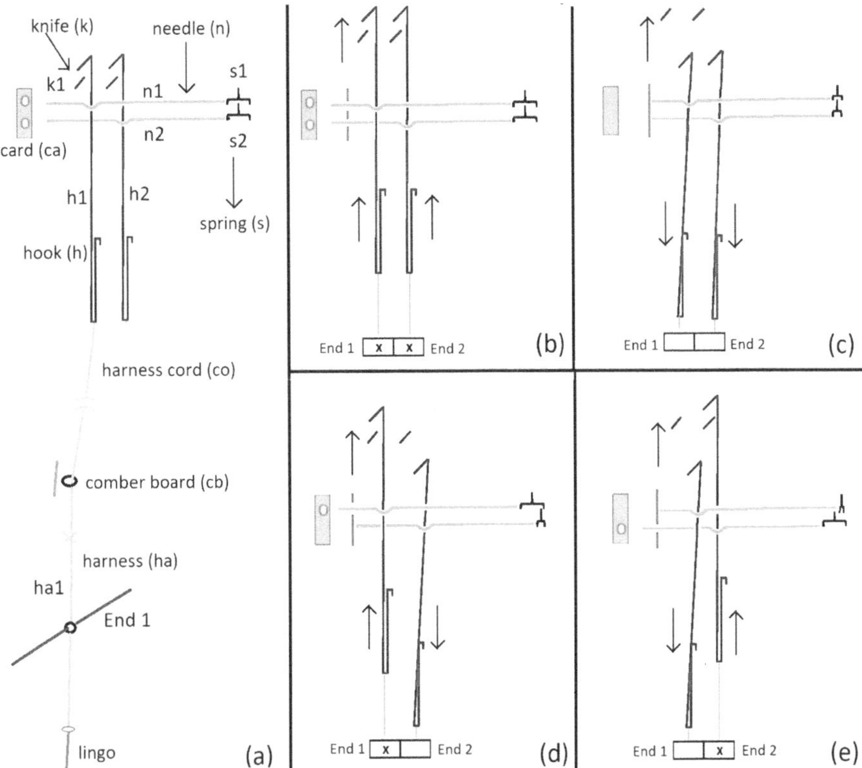

FIGURE 3.1 Working principle of mechanical jacquard: (a) basic parts of jacquard involved in lifting the end, (b) ends are lifted when the card with holes is pressed against the needles, (c) ends are left down when the blank card is pressed against the needles, (d) end 1 is up and end 2 is down when the card with one hole and one blank is pressed against the needles, and (e) end 1 is down and end 2 is up when the card with one blank and one hole is pressed against the needles.

In Figure 3.1a, each needle is connected to one hook and each hook controls one end in one repeat. There is one griffe with knives to lift all the hooks and one cylinder to carry the cards. Hence, this jacquard is called a 'single lift single cylinder (SLSC) jacquard'. In the SLSC jacquard, the capacity of the jacquard is denoted by the number of hooks it has, which is equal to the number of needles and the number of ends per repeat. The number of hooks used to control the ends in the warp for making the figure is denoted as 'figuring capacity'. For example, the 240-hook jacquard has 240 needles connected to it. The figuring capacity of this jacquard is 240 if all 240 hooks are used to control the warp. For this jacquard, the figured graph design is prepared by taking 240 ends in the graph.

Figure 3.1a shows two hooks h1 and h2 operated by two needles n1 and n2. Springs s1 and s2 press the needles forward. There are knives k1 and k2 for lifting the hooks. These two needles, hooks, and knives are for lifting and lowering two ends (ends 1 and 2).

The next point to know is which hooks are to be raised and which are to be left as is. For this, a graph design (pattern) is prepared in which a mark indicates the lifting of the end over the pick and a blank indicates the end down. Then cards are punched for the graph design. For the marks on the graph, holes are punched on the cards, and for the blanks on the graph, the card is left blank without punching the holes.

The cylinder performs the movement like a pendulum, with punched cards, towards the points of the needles. When the punched card is pressed against the needles, the hole in the card does not impart movement to the needle. Hence, the corresponding hook remains in its vertical position on the corresponding knife bar, and upon lifting the knife, the hook is raised. Alternatively, with the blank on the card, push the needle back towards the springboard. This motion forces back the corresponding hook, pushing it away from the knife bar above, and upon lifting the knife, the hook remains stationary.

Consider that it is required to lift both ends 1 and 2. For this, the graph design is prepared with marks at the first and second ends. The card is punched with two holes, corresponding to the two needles. When the cylinder is presented with this card, the two holes in the card allow both the needles n1 and n2 to enter into the card holes and cylinder holes. Hence, the corresponding hooks h1 and h2 remain in their vertical positions on the corresponding knife bars k1 and k2. Upon lifting these knives, both hooks are raised, thereby lifting both ends 1 and 2. This action is shown in Figure 3.1b.

Consider that it is required to keep ends 1 and 2 down. For this, the graph design is prepared with blanks at the first and second ends. The card is not punched with any holes corresponding to the two needles. When the cylinder is presented with this card, the blank parts of the card push back both the needles n1 and n2 and the springs s1 and s2. Hence, the corresponding hooks h1 and h2 move away from their vertical position from the corresponding knife bars k1 and k2. Upon lifting these knives, both hooks stay away and remain stationary. Thus, both ends 1 and 2 remain down. This action is shown in Figure 3.1c.

Consider that it is required to lift end 1 and leave down end 2 (1 up, 1 down). For this, the graph design is prepared with a mark at the first end and a blank at the second end. The card is punched with one hole and one blank, corresponding to the two needles. When the cylinder is presented with this card, the first hole in the card allows the needle n1 to enter into the card hole and cylinder hole. The next blank in the card pushes back the needle n2, by which the spring s2 is also pressed back. Hence, the corresponding hook h1 remains in its vertical position on the corresponding knife bar k1 and the corresponding hook h2 moves away from its vertical position from the corresponding knife bar k2. Upon lifting these knives, hook h1 is raised, thereby lifting end 1, and hook h2 stays away and remains stationary. Thereby, end 1 is lifted and end 2 remains down. This action is shown in Figure 3.1d.

Consider that it is required to leave down end 1 and lift end 2 (1 down, 1 up). For this, the graph design is prepared with a blank at the first end and a mark at the second end. The card is punched with one blank and one hole corresponding to the two needles. When the cylinder is presented with this card, the first blank in the card pushes back the needle n1 and the spring s1 and the next hole in the card allows the needle n2 to enter into the card hole and cylinder hole. Hence, the corresponding

TABLE 3.1

The Comparative Working Principle of Mechanical Jacquard for Lifting the End and Lowering the End

• End up	• End down
• Mark on the graph	• Blank on the graph
• Hole on the card	• No hole (blank) on the card
• Needle is not pressed back	• Needle is pressed back
• Hook remains vertically above the knife bar	• Hook moves away from the knife bar
• Knife moves up	• Knife moves up
• Hook is lifted up	• Hook is not lifted up
• Harness connection lifted up	• Harness connection not lifted up
• End up	• End down

hook h1 moves away from its vertical position on the corresponding knife bar k1, and the corresponding hook h2 remains in its vertical position on the corresponding knife bar k2. Upon lifting these knives, hook h1 stays away and remains stationary, and hook h2 is raised. Thereby, end 1 remains down and end 2 is lifted. This action is shown in Figure 3.1e.

The knife bars are fixed in a griffe. The lifting of the griffe lifts the knives. The knives lift the selected hooks and the harness cords connected to them, thus lifting the ends to form a shed with those not lifted. The connections between the griffe and the cylinder convert the up-and-down movement of the griffe into lateral movement and rotation of the cylinder. After inserting the pick through the shed, the griffe is left down. All the hooks come down to their original positions, and the ends move down to the bottom of the reed by the dead weight (lingo) connected at the bottom of each harness. In the handloom, the lifting of the griffe is performed by the weaver through the treadle whereas it is done by motor power in a power loom. Table 3.1 shows the comparative working principle of jacquard in short.

3.2　JACQUARD CAPACITY AND NEEDLES–HOOKS NUMBERING

The total number of hooks and needles in the jacquard is arranged in a definite number of rows, with a definite number of hooks in each row. In the 240-hook jacquard, the 240 hooks and needles are arranged in 30 short rows with eight hooks per short row (30 × 8). It can also be denoted as eight long rows with 30 hooks per long row (8 × 30). Apart from the regular hooks, some jacquards also have extra hooks and needles for other purposes like controlling the selvedge ends, controlling some mechanisms in weaving, etc. Table 3.2 gives the arrangement of hooks and needles in a few mechanical jacquards widely used.

The jacquard is mounted on the top of the loom, normally 5–6 feet above the warp sheet. It is mounted by keeping the cylinder on any one of the four sides, viz., left, right, front, or back, as per the requirement. The position of the cylinder, numbering of needles and hooks, and all other observations of the jacquard are done from the weavers' view, standing in front of the loom.

TABLE 3.2

Short Row-Wise and Long Row-Wise Arrangements of Hooks and Needles in a Few Mechanical Jacquards

S. No.	Figuring Capacity (Hooks & Needles)	No. of Short Rows (No. of Hooks per Long Row)	No. of Long Rows (No. of Hooks per Short Row)	Extra Hooks & Needles	Extra Hooks & Needles	Total Hooks & Needles
1	120	20	6			120
2	200	25	8	8	12	220
3	240	30	8			240
4	400	50	8	8	12	420
5	600	50	12	12	20	632

The mechanism of the jacquard is routed from a hole or blank space of the card to the needle, hook, and end. Hence, the total needles, hooks, space of the card for holes and blanks, and warp ends are given numbers for counting and connecting them systematically. The numbering is started by knowing the position of the cylinder after mounting the jacquard on the loom. The number is given either short row-wise or long row-wise. The numbering starts with a needle. Then, give the same number to the space on the card that corresponds to the needle, the hook that is controlled by the needle, and the end that is controlled by the hook.

Figure 3.2a–c shows *the needles and hooks of 200 hooks jacquard keeping the cylinder on the left side and following the short row-wise numbering. It is shortly denoted as 'Left side cylinder, short row-wise numbering'.* Figure 3.2a is the front side view of 200 hooks and needles jacquard, showing the first short row with eight hooks and eight needles. There are eight knife bars in the griffe, one for each long row of hooks. The cylinder carrying the card is on the left side. In each short row, the eight needles from top to bottom are connected to eight hooks from left to right. For short row-wise numbering, take the fact that there are 25 short rows and each short row has eight hooks and eight needles. In the first short row, the top needle is given the number 1. This top needle is connected to the left-side hook. Hence, the left-side hook is hook 1. The harness cord, the harness connected, and the end passed through that harness are numbered 1. In the first short row, the eight needles are counted from top to bottom and numbered from 1 to 8. Consequently, the eight hooks are counted from left to right and numbered from 1 to 8.

Figure 3.2b is the right-side isometric view of the same jacquard, showing both the first short row and the last short row (the 25th row) with the first and last needles and hooks in these two short rows. The second short row is at the back side of the first short row. Again, in the second short row, the eight needles are counted from top to bottom (needles 9–16) and the eight hooks are counted from left to right (hooks 9–16). Likewise, the numbering goes for all 25 short rows. Accordingly, in the last short row, the bottom needle is the last needle and the right-side hook is the last hook of the jacquard.

Figure 3.2b also shows the punched card placed facing the needles horizontally over the horizontal cylinder. The space of the card facing the top needle 1 is the space for a mark or blank to operate needle 1, and hence it is marked as 1. Similarly, the

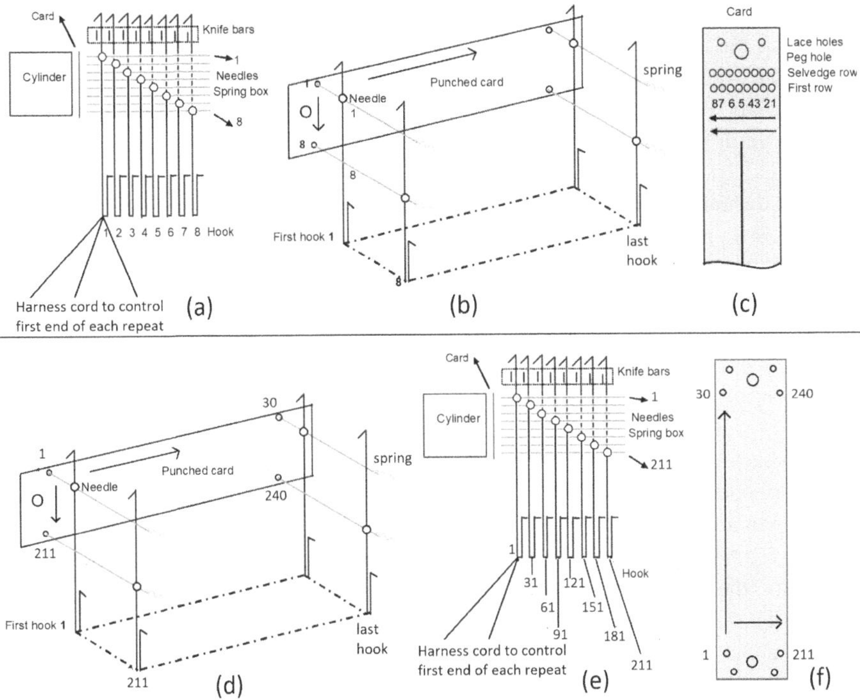

FIGURE 3.2 Needles, card space, hooks, and ends numbering: (a) the first short row of 200 hooks jacquard with the cylinder on the left-side and short row-wise numbering, (b) the first short row and the last row of 200 hooks jacquard with the card facing the needles, (c) 200 hooks jacquard card with short row-wise space numbering indicated by arrows, (d) the first short row and the last row of 240 hooks jacquard with the card facing the needles, (e) the first short row of 240 hooks jacquard with the cylinder on the left-side and long row-wise numbering, and (f) 240 hooks jacquard card with long row-wise space numbering indicated by arrows.

space of the card facing the bottom needle 8 is the space for a mark or blank to operate needle 8, and hence it is marked as 8. In the last row, the bottom-most space is the space for the last needle. The card taken out of the cylinder is shown in Figure 3.2c.

Consolidating the above, it is given that, keeping the cylinder on the left side, in the short row-wise needles and hooks numbering,

i. in the jacquard, the short rows of hooks are counted from front to back, and in each short row of hooks, the hooks are counted from left to right.
ii. in the jacquard, the short rows of needles are counted from front to back, and in each short row of needles, the needles are counted from top to bottom.

Corresponding to the short row-wise needle and hook numbering, the space numbering on the card is also short row-wise (width-wise), *keeping the card vertical*. The counting of short rows (width-wise) is from top to bottom of the card, and in each short row, the counting of space is from right to left. Furthermore, this space numbering for the card is followed while punching the card. While punching, the card is punched short row-wise (width-wise) from top to bottom of the card, and in each short row, the holes are punched from right to left, as indicated by the arrows on the card shown in Figure 3.2c.

Figure 3.2d–f shows *the needles and hooks of 240 hooks jacquard keeping the cylinder on the left side and following the long row-wise numbering. It is shortly denoted as 'Left side cylinder, long row-wise numbering'.* Figure 3.2d is the right-side isometric view of the jacquard, showing both the first and last short rows. Since long row-wise numbering is followed, take the fact that there are eight long rows and each long row has 30 hooks. In the first short row, the top needle is given the number 1. This top needle is connected to the left-side hook. Hence, the left-side hook is hook 1. The harness cord, harness connected, and the end passed through the harness are numbered 1. The second short row is at the back side of the first short row. In the second short row, the top needle is given the number 2. This top needle is connected to the left-side hook. Hence, the left-side hook is hook 2. Similarly, in the third short row, the top needle is given the number 3, and so on. Finally, in the last thirtieth short row, the top needle is given the number 30 and the left side hook is given the number 30. Again, the numbering starts with the second needle in the first short row, which is given the number 31, and the second hook in the first short row is given the number 31. In the last thirtieth short row, the second needle is given the number 60 and the second hook is given the number 60. Accordingly, in the first short row, the bottom needle is the 211th needle and the right-side hook is the 211th hook of the jacquard. In the last short row, the bottom needle is the 240th needle (last) and the right-side hook is the 240th hook (last) of the jacquard.

Figure 3.2e is the front side view of the same jacquard, showing the first short row with eight hooks and eight needles. The cylinder carrying the card is on the left side. In each short row, the eight needles from top to bottom are connected to eight hooks from left to right. There are eight knife bars in the griffe, one for each long row of hooks. All these are similar to the first short row shown in Figure 3.2a. As stated earlier, in the first short row, the top needle is given the number 1. This top needle is connected to the left-side hook. Hence, the left-side hook is hook 1. The harness cord, harness connected, and end passed through the harness are numbered 1. In the first short row, the second needle, which is the first needle in the second long row, is the 31st needle and the second hook connected to this needle, which is the first hook in the second long row, is the 31st hook. Similarly, the third needle and hook become 61, and so on. Accordingly, the last needle and hook in the first short row become 211.

Figure 3.2d also shows the punched card placed facing the needles horizontally over the horizontal cylinder. The space of the card facing the top needle 1 is the space for a mark or blank to operate needle 1, and hence it is marked as 1. Similarly, the space of the card facing the top needle of the last short row (30th needle) is the space for a mark or blank to operate the needle 30, and hence it is marked as 30. Similarly, the space of the card facing the bottom needle of the first short row (211th needle) is the space for a mark or blank to operate needle 211, and hence it is marked as 211. In the last row, the bottom-most space is the space for the last needle (240). The card taken out of the cylinder is shown in Figure 3.2f.

Consolidating the above, it is given that, keeping the cylinder on the left side, in the long row-wise needles and hooks numbering,

i. in the jacquard, the long rows of hooks are counted from left to right, and in each long row, the hooks are counted from front to back.
ii. in the jacquard, the long rows of needles are counted from top to bottom, and in each long row of needles, the needles are counted from front to back.

Corresponding to the long row-wise needle and hook numbering, the space numbering in the card is also long row-wise (length-wise), keeping the card horizontal. The counting of long rows (length-wise) is from top to bottom of the card, and in each long row, the counting of spaces is from left to right. Furthermore, this space numbering is followed while punching the card. While punching, the card is punched long row-wise (length-wise) from top to bottom of the card, and in each long row, the holes are punched from left to right as shown by the arrows on the card.

3.3 HARNESS BUILDING AND CALCULATIONS

The systematic way of connecting all the hooks of the jacquard to all the harnesses of the warp sheet with harness cord by passing it through the holes of the comber board is known as 'harness building'. The system of harness building depends upon (i) the left or right or front or back position of the card cylinder of the jacquard mounted on the loom, (ii) the short row-wise or long row-wise numbering of the needles and hooks, and (iii) the particulars of the jacquard used and the warp sheet. The position of the cylinder and the short row-wise or long row-wise numbering of needles and hooks depend upon the working convenience of the jacquard in the loom and the type of card-punching machine available to the designer for punching.

After understanding the working principle of jacquard and the numbering of hooks and needles, it is essential to understand the comber board through which the harness cords are passed to connect the hooks and harnesses during harness building. The comber board is a plate with holes to pass the harness cords. The comber board can be compared with the reed used for the warp. The reed is to keep the warp at the required width by drawing all the ends of the warp through the dents of the reed. Similarly, the comber board is to keep the harness cords at the required width by drawing all the harness cords through the holes of the comber board. Since all the ends are passed through the harness cords, the width of the warp sheet on the beam, the width of the warp sheet in the reed, and the width of the harness cords in the comber board must be equal. Furthermore, if one hook is connected to one end by one harness cord, the harness cords per unit space in the comber board must be equal to the ends per unit space in the warp.

Similar to the needles and hooks of the jacquard, the holes in the comber board are also in a definite number of rows, with a definite number of holes in each row. The number of holes per row and the number of rows in the comber board depend upon the short row-wise or long row-wise numbering of hooks and the number of hooks in each short row or long row. For example, consider that 200 hooks jacquard (25 × 8) are used for harness building, keeping the cylinder on the left side with short row-wise numbering (8–8) as shown in Figure 3.2a–c. For this, the comber board used must have eight holes per row in length and 25 rows in width (8 × 25 = 200 holes) for one repeat to accommodate 200 harness cords for connecting 200 hooks and 200 harnesses carrying 200 ends. To quote another example, consider that 240 hooks jacquard (30 × 8) are used for harness building, keeping the cylinder on the left side with long row-wise numbering (30–30) as shown in Figure 3.2d–f. For this, the comber board used must have 30 holes per row in length and eight rows in width (30 × 8 = 240 holes) for one repeat to accommodate 240 harness cords for connecting 240 hooks and 240 harnesses carrying 240 ends.

The next step in harness building is to calculate the particulars of the warp for which the harness building has to be carried out and the particulars of the jacquard used for the harness building. The three particulars that are to be noted for carrying the harness construction are: (i) Jacquard figuring capacity; (ii) Ends per inch in the reed; and (iii) Width of warp sheet in the reed. From these particulars, the other particulars calculated are: (iv) Total ends in warp; (v) Ends per repeat; (vi) Total number of repeats; (vii) Width of the repeat (harness-ends); (viii) Number of harness cords per bunch (per hook); (ix) Total number of bunches; and (x) Total harnesses. Finally, the total harnesses must be equal to the total ends in the warp if one harness controls one end in the warp.

The calculation is as follows:

    i. Jacquard figuring capacity is known
    ii. Ends per inch in the reed is known
    iii. Width of warp sheet in the reed is known
    iv. Total ends in warp = Ends per inch in the reed × Width of warp sheet in the reed
    v. Ends per repeat = Jacquard figuring capacity (if one harness controls one end)
    vi. Total number of repeats = $\dfrac{\text{Total ends in warp}}{\text{Ends per repeat}}$
    vii. Width of repeat (harness – ends) = $\dfrac{\text{Width of warp sheet in the reed}}{\text{Total number of repeats}}$

        OR

    vi. Width of repeat (harness – ends) = $\dfrac{\text{Jacquard figuring capacity}}{\text{Ends per inch in reed}}$
    vii. Total number of repeats = $\dfrac{\text{Width of warp sheet in the reed}}{\text{Width of repeat}}$
    viii. Number harness cords per bunch (per hook) = Total number of repeats
    ix. Total number of bunches = Total number of hooks employed (figuring capacity)
    x. Total harnesses = Total number of bunches × Number of harness cords per bunch

As an example, consider that the harness has to be built in a loom using a 200-hook capacity jacquard to control a warp sheet. The width of the warp sheet is 30 inches with 40 ends per inch in the reed. The jacquard is mounted keeping the cylinder on the left side, and the harness is built by short row-wise numbering as shown in Figure 3.2a. The calculation for this setup is as follows.

The particulars known are:

  i. Jacquard figuring capacity is 200 hooks.
 ii. Ends per inch in the reed is 40.
iii. Width of warp sheet in the reed is 30 inches.

Other particulars calculated are:

 iv. Total ends in warp = Ends per inch in the reed × Width of warp sheet in the reed

$$= 40 \text{ per inch} \times 30'' = 1200$$

 v. Ends per repeat = Jacquard figuring capacity (if one harness controls one end) = 200

 vi. Total number of repeats $= \dfrac{\text{Total ends in warp}}{\text{Ends per repeat}} = \dfrac{1200}{200} = 6$

vii. Width of repeat (harness − ends) $= \dfrac{\text{Width of warp sheet in the reed}}{\text{Total number of repeats}} =$

$\dfrac{30}{6} = 5''$

OR

 vi. Width of repeat (harness − ends) $= \dfrac{\text{Jacquard figuring capacity}}{\text{Ends per inch in reed}} = \dfrac{200}{40} = 5''$

vii. Total number of repeats $= \dfrac{\text{Width of warp sheet in the reed}}{\text{Width of repeat}} = \dfrac{30}{5} = 6$

 viii. Number of harness cords per bunch (per hook) = Total number of repeats = 6
  ix. Total number of bunches = Total number of hooks employed (figuring capacity) = 200
   x. Total harnesses = Total number of bunches × Number of harness cords per bunch

$$= 200 \times 6 = 1200$$

Total harnesses = Total ends = 1200

In the comber board, number of holes in a row (length way) = Number of hooks in a short row = 8

In the comber board, number of rows per repeat (width way) = Number of short rows = 25.

Figure 3.3a shows the sketch of the harness building carried in the loom as per the particulars calculated in the above example 1. In the jacquard, the cylinder is on the left side, the arrow indicates the short row-wise numbering of hooks, and the neck

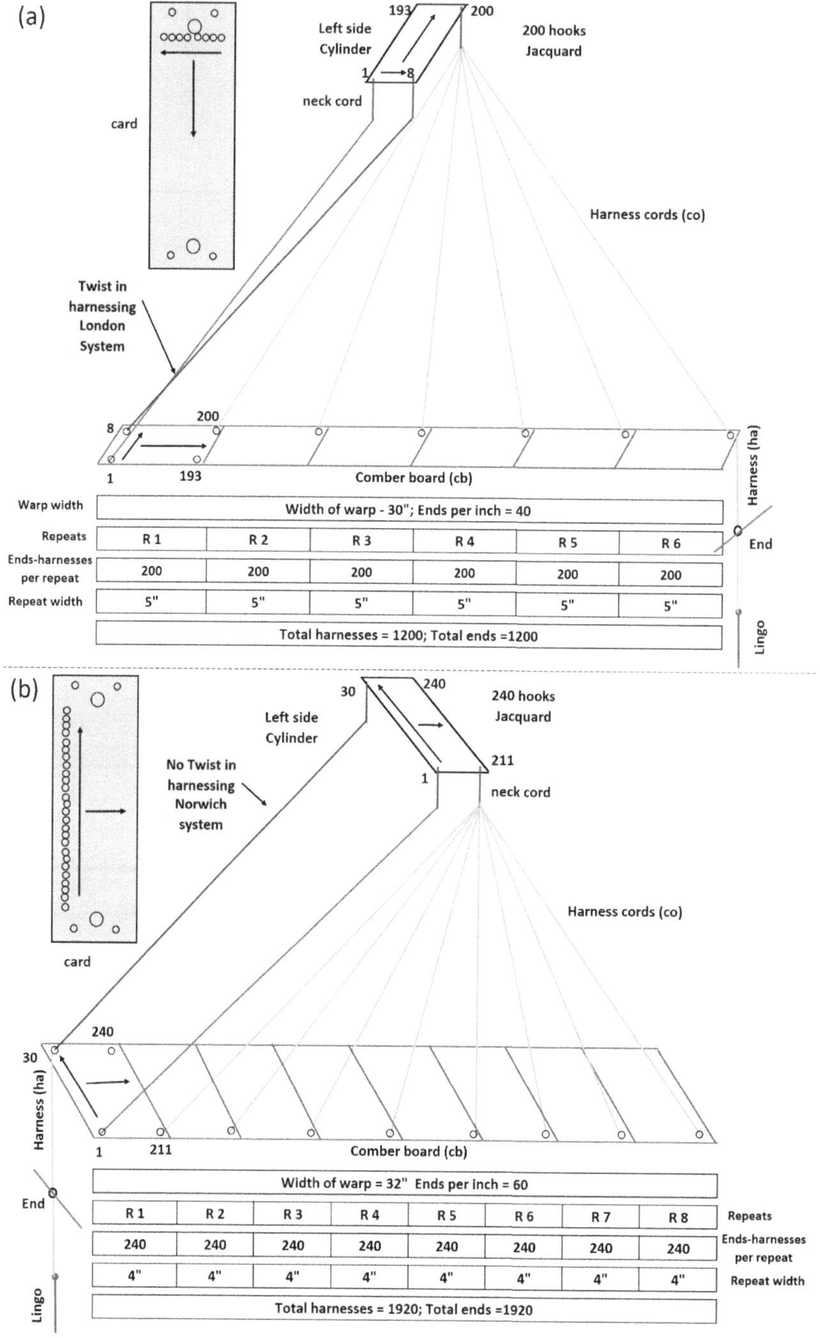

FIGURE 3.3   Two systems of harness building: (a) London system having a twist in harnessing with short row-wise hooks numbering and (b) Norwich system having no twist in harnessing with long row-wise hooks numbering.

cords from hooks 1, 8, and 200 are shown. The full width of the comber board, divided into six parts (six repeats), is shown at the bottom. In each repeat, the arrow shows the length-wise numbering of holes. Holes 1, 8, 193, and 200 are also indicated. The harness cords from hook 1 passed through hole 1 and hook 8 passed through hole 8 are shown. The hooks are numbered from left to right, but the holes on the comber board are numbered from front to back. Hence, while taking the eight harness cords from one short row of eight hooks and passing them through the eight holes, a 90° anti-clockwise turn or twist occurs. This twist occurs on each set of eight cords. This system of harness building is traditionally named the 'London system' and technically noted as 'Quarter twist harness building'. The six harness cords from the last 200th hook pass through the last 200th hole in each repeat. A harness connected to the 200th harness cord is shown at the extreme right. Hence, the end drawn through this harness is the 200th end. The other particulars calculated are indicated at the bottom of the comber board for reference. A full card is also given with arrows showing the method of short row-wise punching.

As another example, consider that the harness has to be built on a loom using a 240-hook capacity jacquard to control the warp sheet. The width of the warp sheet is 32 inches with 60 ends per inch in the reed. The jacquard is mounted by keeping the cylinder on the left side. The harness is built by long row-wise numbering as shown in Figure 3.2d. The calculation for this setup is as follows.

The particulars known are:

   i. Jacquard figuring capacity is 240 hooks.
   i. Ends per inch in the reed is 60.
 iii. Width of warp sheet in the reed is 32 inches.

Other particulars calculated are:

  iv. Total ends in warp = Ends per inch in the reed × Width of warp sheet in the reed

$$= 60 \text{ per inch} \times 32'' = 1920$$

   v. Ends per repeat = Jacquard figuring capacity (if one harness controls one end) = 240

$$\text{vi. Total number of repeats} = \frac{\text{Total ends in warp}}{\text{Ends per repeat}} = \frac{1920}{240} = 8$$

$$\text{vii. Width of repeat (harness} - \text{ends)} = \frac{\text{Width of warp sheet in the reed}}{\text{Total number of repeats}} = \frac{32}{8} = 4''$$

   OR

$$\text{vi. Width of repeat (harness} - \text{ends)} = \frac{\text{Jacquard figuring capacity}}{\text{Ends per inch in reed}} = \frac{240}{60} = 4''$$

$$\text{vii. Total number of repeats} = \frac{\text{Width of warp sheet in the reed}}{\text{Width of repeat}} = \frac{32}{4} = 8$$

viii. Number of harness cords per bunch (per hook) = Total number of repeats = 8

ix. Total number of bunches = Total number of hooks employed (Figuring capacity) = 240

x. Total harnesses = Total number of bunches × Number of harness cords per bunch

$= 240 \times 8 = 1920$

Total harnesses = Total ends = 1920

In the comber board, number of holes in a row (length way) = Number of hooks in a long row = 30

In the comber board, number of rows per repeat (width way) = Number of long rows = 8.

Figure 3.3b shows the sketch of the harness building carried in the loom as per the particulars calculated in the above example 2. In the jacquard, the cylinder is on the left side, the arrow indicates the long row-wise numbering, and the neck cords from hooks 1, 30, and 211 are shown. The full width of the comber board, divided into eight parts (eight repeats), is shown at the bottom. In each repeat, the arrow shows the length-wise numbering of holes. Holes 1, 30, 211, and 240 are also indicated. The harness cords from hook 1 passed through hole 1 and hook 30 passed through hole 30 are given. The hooks are numbered from front to back, and the holes are also numbered from front to back. Hence, while taking the 30 harness cords from one long row of 30 hooks and passing them through the 30 holes of the comber board, all the cords stay parallel without any twist as that of a 90° anti-clockwise turn or twist occurs in the London system. This system of harness building is traditionally named the 'Norwich system' and technically noted as 'parallel harness building'. The eight harness cords from the 211th hook pass through the 211th hole in each repeat of the comber board. Similarly, each harness cord from each hook is passed through the respective holes in the comber board. A harness connected to the 30th harness cord from the 30th hook is shown at the extreme left. Hence, the end drawn through this harness is the 30th end. The other particulars calculated are indicated at the bottom of the comber board for reference. A full card is also indicated with arrows showing the method of long row-wise punching as shown in Figure 3.3b.

Figure 3.4 shows the following different ways of numbering the needles and hooks along with the numbering of holes in the comber board and space on the card according to the jacquard mounting, cylinder side, London, and Norwich harness building systems.

(i) Left-side cylinder with short row-wise numbering (Figure 3.4a), as explained above, (ii) Right-side cylinder with short row-wise numbering (Figure 3.4b), (iii) Left-side cylinder with long row-wise numbering (Figure 3.4c), as explained above, (iv) Right-side cylinder with long row-wise numbering (Figure 3.4d), (v) Front side cylinder with short row-wise numbering (Figure 3.4e), and (vi) Back side cylinder with short row-wise numbering (Figure 3.4f).

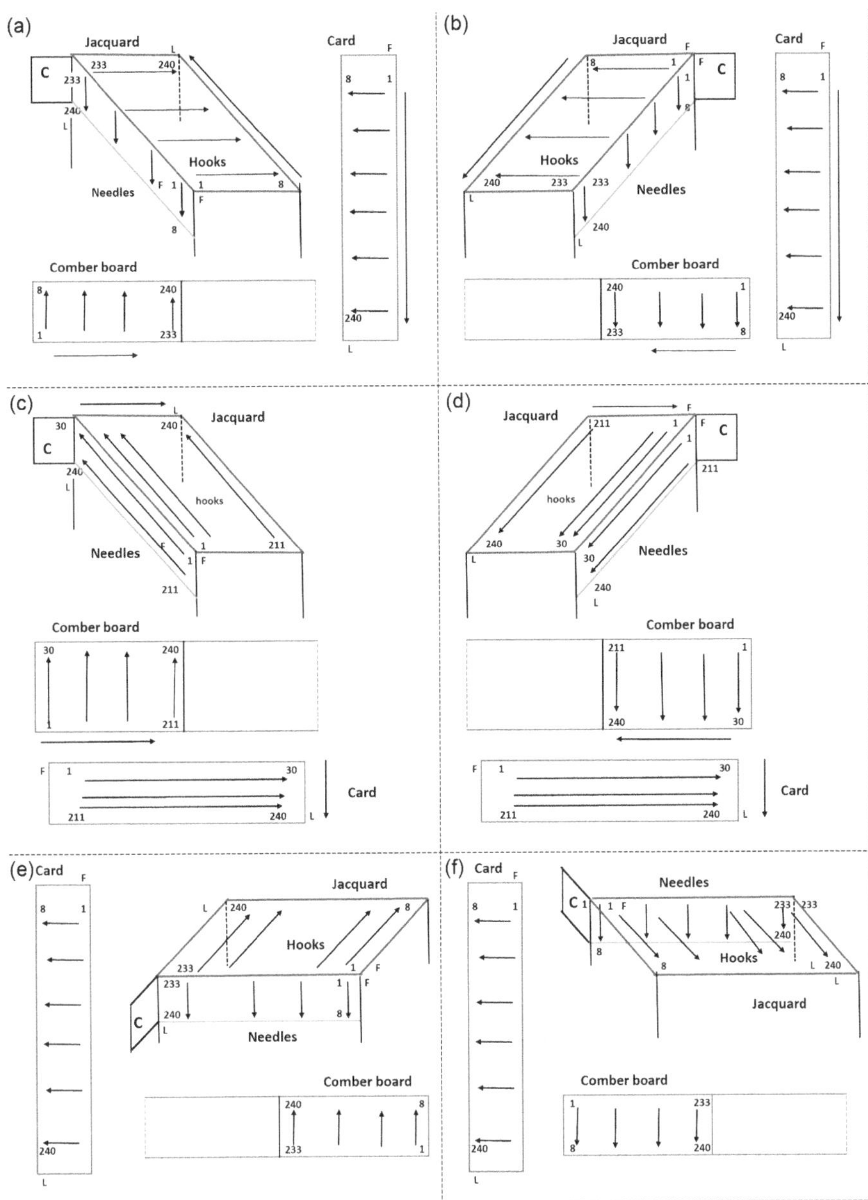

FIGURE 3.4 Different methods of jacquard mounting: (a) left-side cylinder with short row-wise numbering, (b) right-side cylinder with short row-wise numbering, (c) left-side cylinder with long row-wise numbering, (d) right-side cylinder with long row-wise numbering, (e) front side cylinder with short row-wise numbering, and (f) back side cylinder with short row-wise numbering.

By comparing parts of Figure 3.4a and b, it is noted that the jacquard mounted with the left-side cylinder is rotated 180° clockwise to make the jacquard mounted with the right-side cylinder. In both cases, the London system of harness building is followed. Though the directions of needles, hooks, and harness holes numbering change, the direction of holes numbering in the card is same in both the cases. This enables to follow the same short row-wise punching method for the cards used in these two jacquard setups. Thus, the card set punched for the left-side cylinder can be used for the right-side cylinder.

Similarly, by comparing parts of Figure 3.4c and d, it is noted that the jacquard mounted with the left-side cylinder is rotated 180° clockwise to make the jacquard mounted with the right-side cylinder. In both cases, the Norwich system of harness building is followed. Though the directions of needles, hooks, and harness holes numbering change, the direction of holes numbering in the card is same in both the cases. This enables to follow the same long row-wise punching method for the cards used in these two jacquard setups. Thus, the card punched for the left-side cylinder can be used for the right-side cylinder.

Likewise, by comparing parts of Figure 3.4a and e, it is noted that the jacquard mounted with the left-side cylinder is rotated 90° anti-clockwise to make the jacquard mounted with the front side cylinder. In both cases, the London system of harness building is followed. Though the directions of needles, hooks, and harness holes numbering change, the direction of holes numbering in the card is same in both the cases. This enables to follow the same short row-wise punching method for the cards used in these two jacquard setups. Thus, the card punched for the left-side cylinder can be used for the front side cylinder.

Again, by comparing parts of Figure 3.4a and f, it is noted that the jacquard mounted with the left-side cylinder is rotated 90° clockwise to make the jacquard mounted with the back side cylinder. In both cases, the London system of harness building is followed. Though the directions of needles, hooks, and harness holes numbering change, the direction of holes numbering in the card is same in both the cases. This enables to follow the same short row-wise punching method for the cards used in these two jacquard setups. Thus, the card punched for the left-side cylinder can be used for the front side cylinder.

Apart from the above, the hooks numbering in each short row can also be from right to left instead of from left to right and the hooks numbering in each long row can also be from back to front instead of from front to back. Whatever method is followed for counting the hooks and needles in the row and counting the rows, it is important to keep in mind that, according to the needles and hooks numbering, the harness building is to be followed and also the punching of cards.

3.4 ELECTRONIC JACQUARD—WORKING PRINCIPLE

Bonas, Grosse, and Staubli are electronic jacquard manufacturers. The important parts of the Staubli electronic jacquard are shown in Figure 3.5a–f.

The electromagnetic module shown in Figure 3.5a with the necessary mechanisms enclosed in it is the heart of the electronic jacquard. Each module consists of

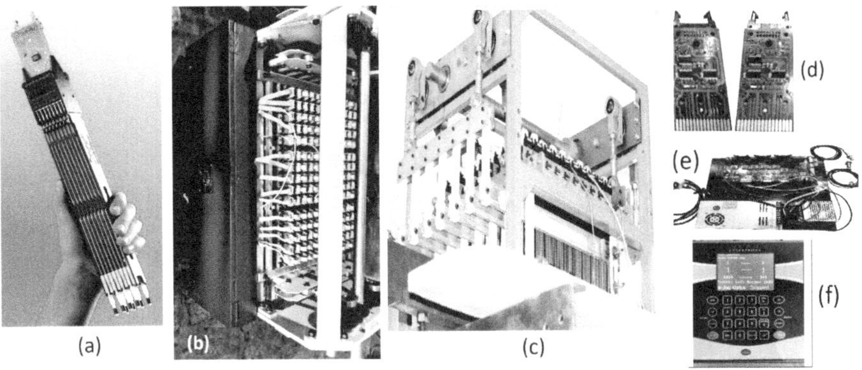

FIGURE 3.5   Different parts of electronic jacquard: (a) the electromagnet module, (b) the arrangement of electromagnets, (c) the mechanical double-lift knives system, (d) the electronic relay board, (e) the electronic distribution system, and (f) the electronic design control box.

eight sets of parts, viz., electromagnets, retaining hooks, movable hooks, pulleys, and couplings to connect the harness bunch. According to the hook's capacity, the required number of magnet modules are arranged in rows and columns as shown in Figure 3.5b. The hook-lifting mechanical mechanism shown in Figure 3.5c works with the double-lift principle. The magnets in the modules are connected to the electronic relay boards (Figure 3.5d), the electronic distribution system (Figure 3.5e), and the electronic design control box (Figure 3.5f). All these electronic systems, the arrangement of modules, and the mechanical double-lifting system enclosed in a specially designed mechanical enclosure put together are called 'electronic jacquard'.

The digital graph is prepared using figured graph designing software and transferred to the design control box. The weaver can operate the required design by changing the display provided in the design control box using buttons. The electronic control box has a microcontroller and internal memory. The digital graph design with marks and blanks is stored in the internal memory. When both knives are at the centre, the microcontroller reads the data (marks and blanks on digital graph) of the design row by row (pick by pick) serially and transmits the data to the distribution box. The distribution box transmits the data to each column of relay boards. From each relay board, the module gets the data. The data generated by the marks of the digital graph switches on the power supply and makes the solenoids magnetize, and the data generated by the blanks of the digital graph does not switch on the power supply and makes the solenoids not magnetize (inactive).

The working principle followed in the Staubli electronic jacquard is shown in Figure 3.6a–h. The basic parts involved in operating an end are given in Figure 3.6a. To control each end of the warp, there are two movable hooks, h1 and h2. Both hooks h1 and h2 carry projections pr1 and pr2, respectively. The two lifting knives, k1 and k2, are placed below the hook's projections, pr1 and pr2, respectively. The knives move up and down diametrically opposite to each other and work for alternate picks. There are two pulleys, pl1 and pl2, fixed one above the other in a link and free to turn

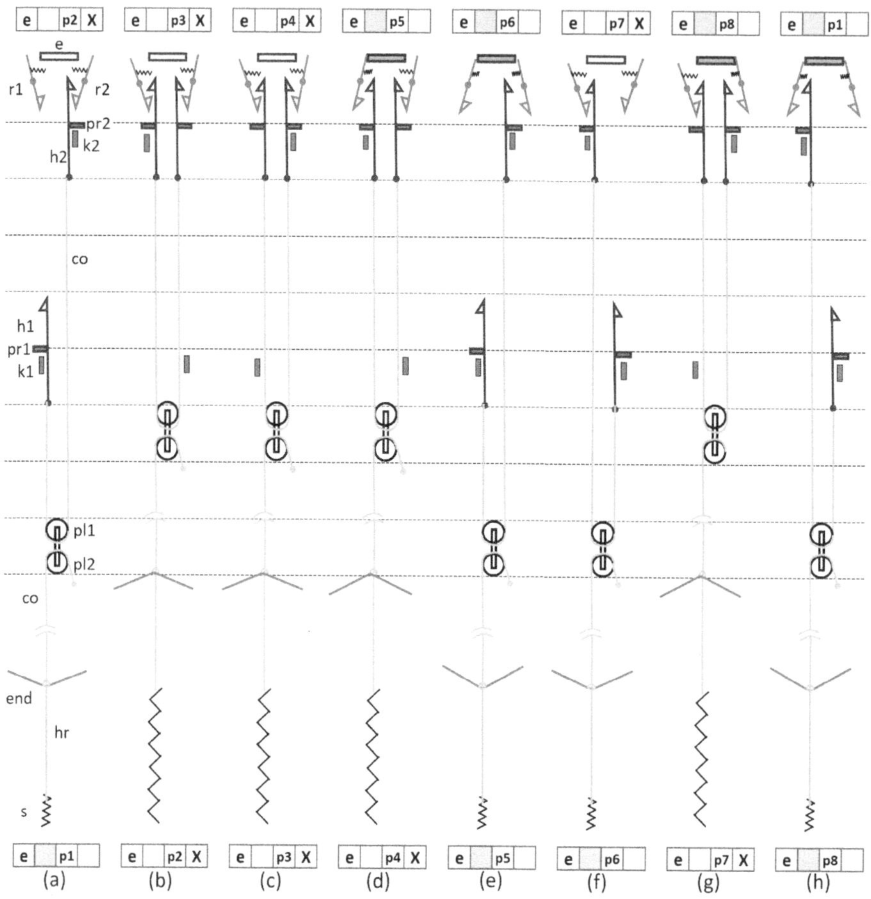

FIGURE 3.6 Working principle of electronic jacquard: (a) the initial position of different parts, (a) and (b) lowering of the end for the first pick and lifting of the end for the second pick (down, up), (c) and (d) lifting of the end for two picks successively (up, up), (e) and (f) lowering of the end for two picks successively (down, down), and (g) and (h) lifting of the end for the first pick and lowering of the end for the second pick (up, down).

on their axes. A cord co1 passes around the pulley pl1, whose ends are connected at the bottom of two hooks. A cord co2 passes around the pulley pl2, whose one end is anchored and the other end is connected to the harness hr, carrying an end in its mail. The bottom of the harness is connected to a spring. On top is an electromagnet (e). The mark in the digital graph activates the electromagnet through the electronic connections. The retaining hooks, r1 and r2, and their fulcrums are at the centre. The top of the retaining hooks faces the electromagnet and the bottom of these hooks faces the projection of the movable hooks. The tops of the retaining hooks are pushed away from the magnets by the springs.

Following are the points that are to be noted in the working of electronic jacquard.

1. A mark on the digital figured graph makes the electromagnet activate whereas a blank in the digital figured graph makes the electromagnet inactive.
2. This activation happens in advance for each pick, that is, the action of the magnets for the second pick happens when the knives are moving to lift or lower the end for the first pick.
3. Every time one of the knives moves to the top, the projection in the movable hook moves the bottom of the retaining hook, thereby causing the top of the retaining hook to move nearer to the electromagnet.
4. If the magnet is magnetized, it catches the top of the retaining hook, thereby moving the bottom of the retaining hook away from the movable hook without locking it. Whereas, if the magnet is not magnetized, it does not catch the top of the retaining hook, thereby keeping the bottom of the retaining hook to lock the movable hook.

The comparative working principle of electronic jacquard for lifting and lowering the end is stated in Table 3.3.

By comparing the working principles of mechanical jacquard given in Table 3.1 and the working principles of electronic jacquard given in Table 3.3, the difference between these two principles could be understood. In mechanical jacquard, for the mark in the graph, finally, the end is lifted up, and for the blank in the graph, the end is left down. On the contrary, in an electronic jacquard, for the mark in the digital graph, the end is lowered down and for the blank in the digital graph, the end is lifted up.

TABLE 3.3

The Comparative Working Principle of Electronic Jacquard for Lifting and Lowering the End

• End up	• End down
• Blank in the digital graph	• Mark in the digital graph
• The electromagnet does not get magnetized	• The electromagnet gets magnetized
• The top of the retaining hook is away from the magnet	• The magnet catches the top of the retaining hook
• The bottom of the retaining hook keeps locking the movable hook	• The bottom of the retaining hook moves away from the movable hook without locking it. Hence, the movable hook rest on the knife
• The knife, which moves up, lifts the movable hook	• The knife, which moves down, brings down the movable hook
• Since one movable hook is locked and other movable hook is also lifted up, it lifts the harness connection	• Since both the movable hooks are resting over the knives, the up-and-down movement of the knives does not lift the harness connection
• End up	• End down

Hence, the designer has to keep this in mind while doing the graph designing for these two jacquards.

The initial position of the end lifting parts while inserting the first pick is shown in Figure 3.6a.

Since the activation of the electromagnet happens in advance for each pick, the activation of the electromagnet for pick 1 is shown at the previous pick, which is the last pick (eighth pick) in the figure explanation.

During this movement of the knives, on the top, the magnetizing of the electromagnet occurs according to the lifting or lowering of the end for pick 1. Consider that, for pick 1, the end is to be lowered down. Hence, there is a colour mark in the digital graph that causes the electromagnet 'e' to magnetize.

This is shown by the rectangle given at the top of Figure 3.6h, where 'e' in the first square and green in the second square show that the electromagnet is magnetized due to the mark in the digital graph. 'p1' in the third square and blank in the fourth square show that for pick 1, the end is down.

This makes the electromagnet catch the top of the retaining hooks r1 and r2 and makes the bottom of the retaining hooks move away from the movable hooks. Hence, when the movable hook h1 is moved to the top, it is not locked with the retaining hook r1, as shown in Figure 3.6h.

Now, the weaver presses the pedal, which makes the knives move. The knife k1 moves down along with the movable hook h1. At the same time, the knife k2 moves up with the movable hook h2. Hence, it makes no movement in the link of pulleys pl1 and pl2, thereby keeping the harness and end without any movement at the bottom, as shown in Figure 3.6a. Now pick 1 is inserted. The rectangle at the bottom of Figure 3.6a is the same as the rectangle shown at the top of Figure 3.6h. This is to show that what was decided has been achieved.

With the completion of pick 1, the knife k1 is at the bottom, along with the movable hook h1. At the same time, the knife k2 is on top, along with the movable hook h2.

During this movement of the knives, on the top, the magnetizing of the electromagnet occurs according to the lifting or lowering of the end for pick 2. Consider that, for pick 2, the end is to be lifted up. Hence, there is a blank in the digital graph, which causes the electromagnet 'e' not to magnetize.

This is shown by the rectangle shown at the top of Figure 3.6a, where 'e' in the first square and blank in the second square show that the electromagnet is not magnetized due to the blank in the digital graph. 'p2' in the third square and mark in the fourth square show that for the pick 2, the end will be lifted up.

This makes the top of the retaining hooks r1 and r2 away from the magnet and the bottom of the retaining hooks are towards the movable hooks. Hence, when the movable hook h2 is moved to the top, it is locked with the retaining hook r2, as shown in Figure 3.6a.

Now, the weaver presses the pedal, which makes the knives move. The knife k1 moves up along with the movable hook h1. At the same time, the knife k2 moves down without the movable hook h2. Since the movable hook h2 was already locked on top, when knife k1 moves up along with the movable hook h1, the link of pulleys pl1 and pl2 also moves up, thereby lifting the harness and end, as shown in Figure 3.6b. Now pick 2 is inserted. The rectangle at the bottom of Figure 3.4b is the same as the rectangle shown at the top of Figure 3.6a. This is to show that what was decided has been achieved.

With the completion of pick 2, the knife k1 has moved to the top along with the movable hook h1. At the same time, the knife k2 has moved down without the movable hook h2 since the movable hook h2 was locked on the top, as shown in Figure 3.6b.

During this movement of the knives, on the top, the magnetizing of the electromagnet occurs according to the lifting or lowering of the end for pick 3. Consider that, for pick 3, the end is to be lifted up. Hence, there is a blank given in the digital graph that causes the electromagnet not to magnetize, which is shown as blank in Figure 3.6b.

This is shown by the rectangle given at the top of Figure 3.6b, where 'e' in the first square and blank in the second square show that the electromagnet is not magnetized due to the blank in the digital graph. 'p3' in the third square and mark in the fourth square show that for the pick 3, the end will be lifted up.

When the knife k1 reaches the top with the movable hook h1, the movable hook h1 moves the bottom of the retaining hook r1 away and the top of it towards the magnet. Since the magnet is not magnetized, it does not catch the top of the retaining hook r1 and keeps the bottom of the retaining hook towards the movable hook h1. Hence, when the movable hook h1 is moved to the top, it is locked with the retaining hook r1, as shown in Figure 3.6b.

Now, the weaver presses the pedal, which makes the knives move. Since the hook h1 was locked on top, the knife k1 moves down without the movable hook h1. Since the hook h2 was also locked on top, the knife k2 moves up without the movable hook h2. Hence, it does not make any movement in the link of pulleys pl1 and pl2, thereby keeping the harness and end without any movement at the top, as shown in Figure 3.6c. Now pick 3 is inserted. The rectangle at the bottom of Figure 3.4c is the same as the rectangle shown at the top of Figure 3.6b. This is to show that what was decided has been achieved.

Similarly, the working principle for other picks 4, 5, 6, 7, and 8 can be understood through Figure 3.6d–h.

3.5 JACQUARD CAPACITY, HOOKS NUMBERING, AND HARNESS BUILDING

The capacity of electronic jacquard used in the industry ranges from 480 to 5376 hooks. The different capacities of electronic jacquards and the common arrangement of eight hook modules in a number of rows and columns are shown in Figure 3.7a. The arrangement of 120 modules (960 hooks) in 10 columns and 12 rows standing at the weaver's seat is shown in Figure 3.7b. 10 columns × 12 rows = 120 modules and 8 hooks per module × 120 modules = 960 hooks. The numbering of modules starts from the front right of the weaver (bottom right in the figure) and goes back. It completes at the back left (top left in the figure), as shown in Figure 3.7b. The method followed

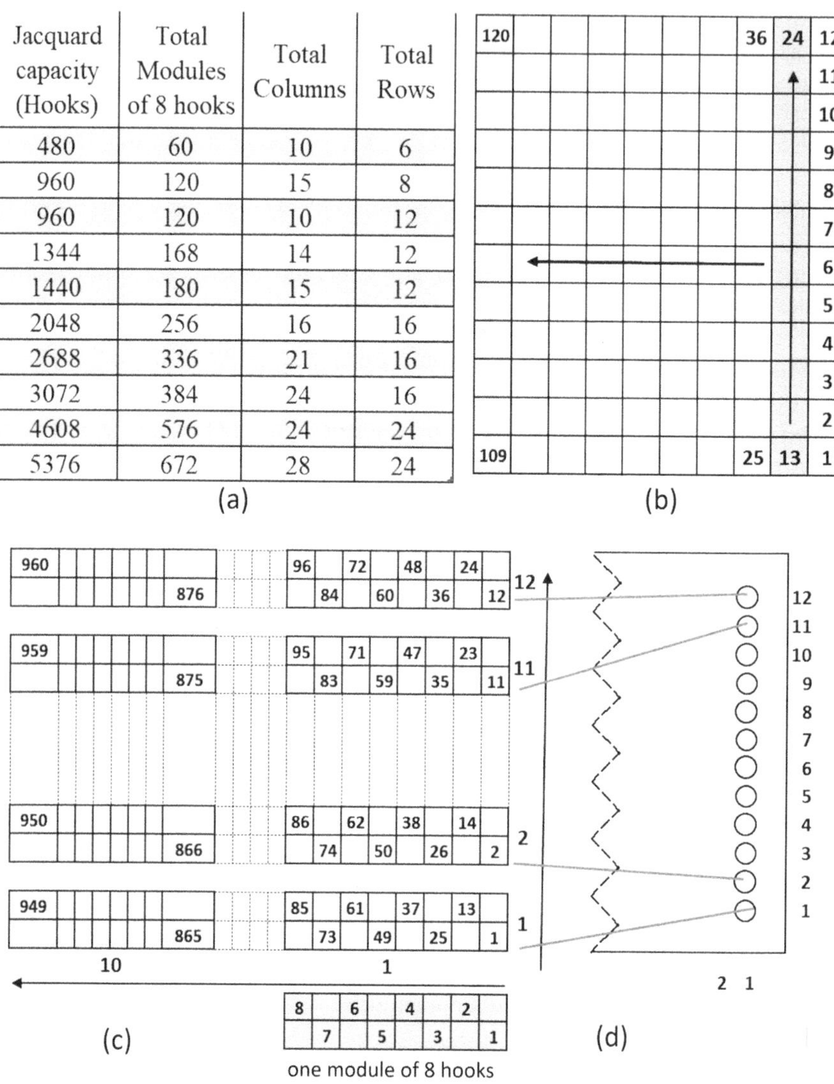

FIGURE 3.7 Jacquard capacity, hooks numbering, and harness building: (a) different capacities of electronic jacquards and the arrangement of modules in rows and columns, (b) arrangement of 120 modules (960 hooks) in 10 columns and 12 rows, (c) numbering of hooks in each module, and (d) harness building from right-side first-row hooks to right-side first-row holes of comber board.

in the numbering of hooks is shown in Figure 3.7c. The numbering of hooks starts by taking one hook from each module, as shown in Figure 3.7c. Hence, it becomes 12 hooks per column and 80 columns of hooks (10 modules per column × 8 hooks per column) totalling to 960 hooks. Figure 3.7d shows the harness building from the right-side first-row hooks to the right-side first-row holes of the comber board.

The harness building shows that the harness cords are passed parallelly from hooks to the holes in the comber board. Hence, it is the Norwich system of harness building (parallel harnessing).

BIBLIOGRAPHY

1. Donaldson, A. (1993). A new era in weaving design. *Canadian Textile Manual.* 7/8, 28–32.
2. Grosicki, Z. J. (2004). Elements of Jacquard shedding. In *Watson's Textile Design and Colour* (pp. 185–191). Woodhead Publishing Limited.
3. Kienbaum, M. (1994). Jacquard technology with electronic control. ITB. *Yarn and Fabric Forming.* 2, 49–58.
4. Kohlhas. (1991). Basic principles of electronically controlled jacquard. *Melliand Textilber.* 72, 168–169.
5. Talukdar, M.K. Sriramulu, P.K., and Ajgaonkar, D.B. (1998). Jacquard shedding. In *Weaving Machines, Mechanisms, Management* (pp. 261–273). Mahajan Publishers Pvt Ltd.

4 Algorithm for Manual and Computer-Aided Figured Graph Designing

4.1 ALGORITHM OF MANUAL FIGURED GRAPH DESIGNING

Before learning Computer-Aided Figured Graph Designing (CAFGD), the designers can practise how to do the figured graphing manually. This practice gives a clear understanding of the algorithm of CAFGD. The flowchart shown in Figure 4.1 is the general algorithm followed for preparing the figured graph manually for producing a figured single cloth. Figure 4.2 shows the different stages of preparing a figured graph manually.

4.1.1 DECIDING THE MOTIF SIZE, GRAPH SIZE, GRAPH COUNT, AND WEAVES

Let us assume that the motif size is 1-inch width by 1-inch length. The ends and picks of the figured graph to be prepared for this motif are 48 ends × 48 picks. The count of the graph sheet is 10 × 10. The size of the 48 ends and 48 picks on the 10 × 10 graph count is 4.8″ × 4.8″. The weave applied on the figure is 8-thread satin and on the ground is 8-thread sateen.

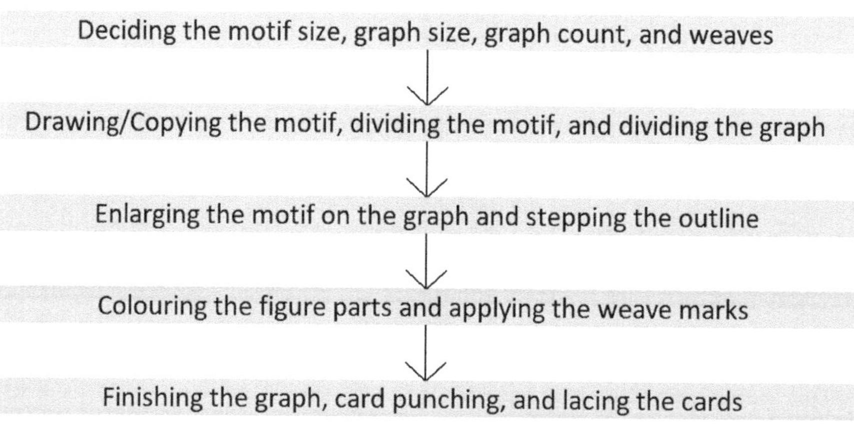

FIGURE 4.1 Flowchart showing the algorithm for manual figured graph designing.

DOI: 10.1201/9781003441205-4

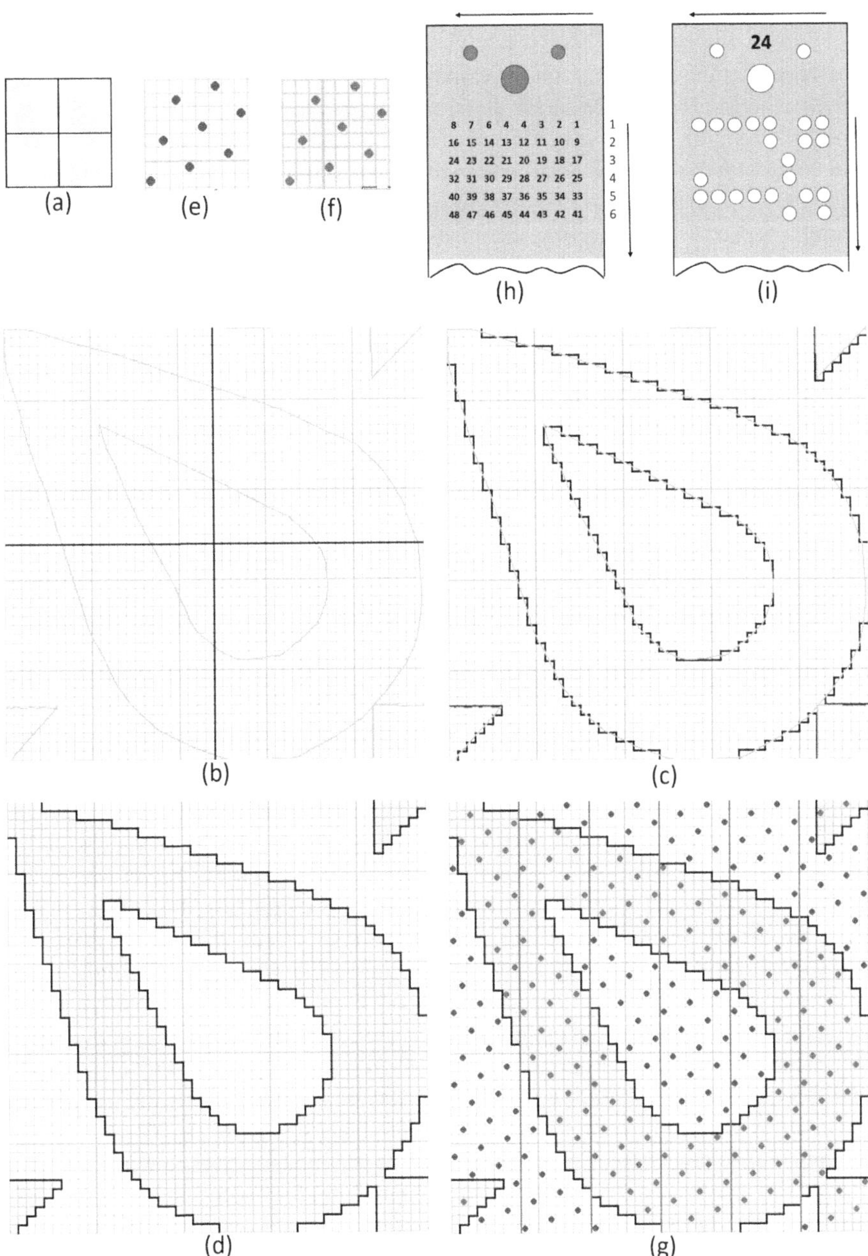

FIGURE 4.2 Different stages of manual figured graph designing: (a) drawing the motif, (b) enlarging the motif outline, (c) stepping the outline, (d) colouring the figure part, (e) weave for the ground, (f) weave for the figure parts, (g) inserting the weave mark in the ground and figure parts, (h) jacquard card to operate eight needles/hooks per short row, and (i) punched card as per the 24th pick of the graph.

4.1.2 Drawing or Copying the Motif and Dividing the Motif and Graph

The motif for the figured graph development may be drawn from scratch or copied from the books. Figure 4.2a shows the figured motif drawn on 1-inch width by 1-inch length. The motif and the graph paper are divided into the same number of vertical and horizontal divisions to enlarge the motif on the graph paper perfectly. Thus, the 1-inch motif is divided into two divisions, each of 0.5 inch, as shown in Figure 4.2a. A graph paper of 48 ends × 48 picks is taken. It is also divided into two vertical and horizontal divisions, each of 24 threads, as shown in Figure 4.2b.

4.1.3 Enlarging the Motif on the Graph Paper and Stepping the Outline

The next step is to transfer the outline of the motif (1″ × 1″ size) onto the graph paper (4.8″ × 4.8″ size, 48 ends × 48 picks) by enlarging it. The figure outline is proportionately enlarged and drawn on the graph by following the division lines in the motif and the graph paper. Figure 4.2b shows the point paper with an enlarged outline of the figure. But the curved outlines thus enlarged on the graph paper pass through the grid lines of the graph. The outline is made to pass through the nearest vertical and horizontal grid lines by converting them into a stepped outline. The stepped outline of the curved outline is shown in Figure 4.2c.

4.1.4 Colouring the Figure Parts and Applying the Weave Marks

Once the stepping of outlines is completed, the figure parts are painted using a transparent colour by leaving the ground white, as shown in Figure 4.2d. The painted figure indicates warp-up (ends up), and the white ground indicates weft-up (ends down). Now, it is necessary to control or bind the floats in the figure parts and ground. Let us assume that the binding weave marks for the ground are 8-thread sateen and the figure is 8-thread satin (sateen marks over painted colour). The one repeat of these two weaves is shown in Figure 4.2e and f, respectively. The weave marks are applied using another colour over the painted figure and ground to control the floats. In manual graph designing, the designer decides the type of weave marks applied for controlling the floats and applies them to the selected parts. While applying the weave marks, the designer maintains the perfection of the boundary by avoiding the marks wherever they are not required as per the cloth variety. After the insertion of the binding weave, the painted figure (yellow) indicates ends up, and the float control binding weave marks (red) over the painted figure indicate ends down. The white ground indicates ends down, and the float control binding weave marks (blue) over the white ground indicate ends up. The figured graph, completed with the necessary binding weave marks, is shown in Figure 4.2g.

4.1.5 Finishing the Graph, Punching, and Lacing the Cards

The final work in preparing the figured graph design, for card punching, is to divide the ends into segments by drawing the vertical lines so that it contains a group of ends according to the number of hooks/needles in the jacquard short row/long row.

In the figured graph shown in Figure 4.2g, each segment is divided into eight ends to suit eight hooks/needles in a short row of the jacquard. The finished graph paper is now ready for punching. One card is punched by reading each pick on the figured graph paper. The punching procedure to be followed is—holes are punched for the yellow painted squares (ends up) except for the red binding weave marks (ends down) over the painted figure. Also, holes are punched for the blue weave marks (ends up) over the white ground (ends down). Figure 4.2h shows the top part of a card having space to punch eight holes in its width to suit a jacquard machine with eight hooks/ needles in a short row. While punching, in each pick, the ends are read from right to left. The numbers given on the card in each row from right to left indicate the numbering of holes. It corresponds to the reading of ends in each pick of the graph from right to left. Figure 4.2i shows the punched card of the 24th pick. Each card is given the same number as the pick number. After completing the punching, the cards are laced in sequence to make a chain and then taken to mount on the jacquard for weaving.

4.2 ALGORITHM FOR COMPUTER-AIDED FIGURED GRAPH DESIGNING

The flowchart shown in Figure 4.3 is the general algorithm for preparing the figured graph using any CAFGD software. Figure 4.4 shows the different stages of preparing a figured graph using the CAFGD software.

4.2.1 Deciding the Motif Size, Graph Size, Graph Count, and Weaves

Let us assume that the motif size is 1-inch width by 1-inch length. The ends and picks of the figured graph to be prepared for this motif are 48 ends × 48 picks. The count of the graph sheet is 10 × 10. A motif of 1-inch × 1-inch is shown in Figure 4.4a. Its four repeats are shown in Figure 4.4b. Then, the weaves to be applied on the different figure parts and ground are decided. Let us consider that the 8-thread sateen weave is for the ground part (outside the figure), and the one-up, seven-down twill weave is also for the ground part (inside the figure), as shown in Figure 4.4c and d, respectively. The 8-thread satin weave is for the figure part, as shown in Figure 4.4e. In these weaves, each colour indicates either end-up or end-down, as detailed below.

   i. In the sateen weave shown in Figure 4.4c, the yellow colour indicates ends down, and the blue colour weave marks indicate ends up.
  ii. In the twill weave shown in Figure 4.4d, the brown colour indicates ends down, and the orange colour weave marks indicate ends up.
 iii. In the satin weave shown in Figure 4.4e, the red colour indicates ends up, and the green colour weave marks indicate ends down.

4.2.2 Drawing the New Motif

There are two ways to draw the new motif. (i) The designer can create the new motif directly on the computer. For this, a new graph file is opened in the required size

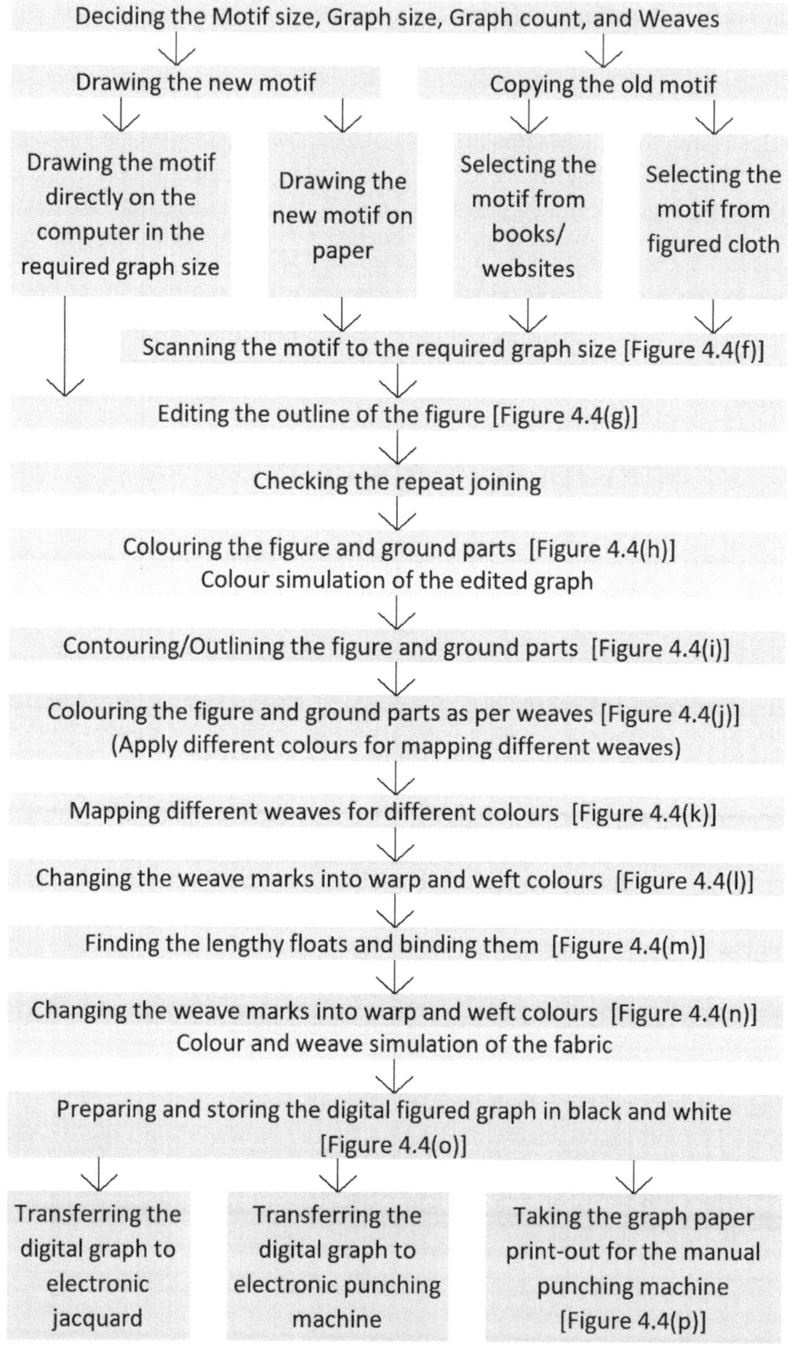

FIGURE 4.3   Flowchart showing the algorithm for computer-aided figured graph designing.

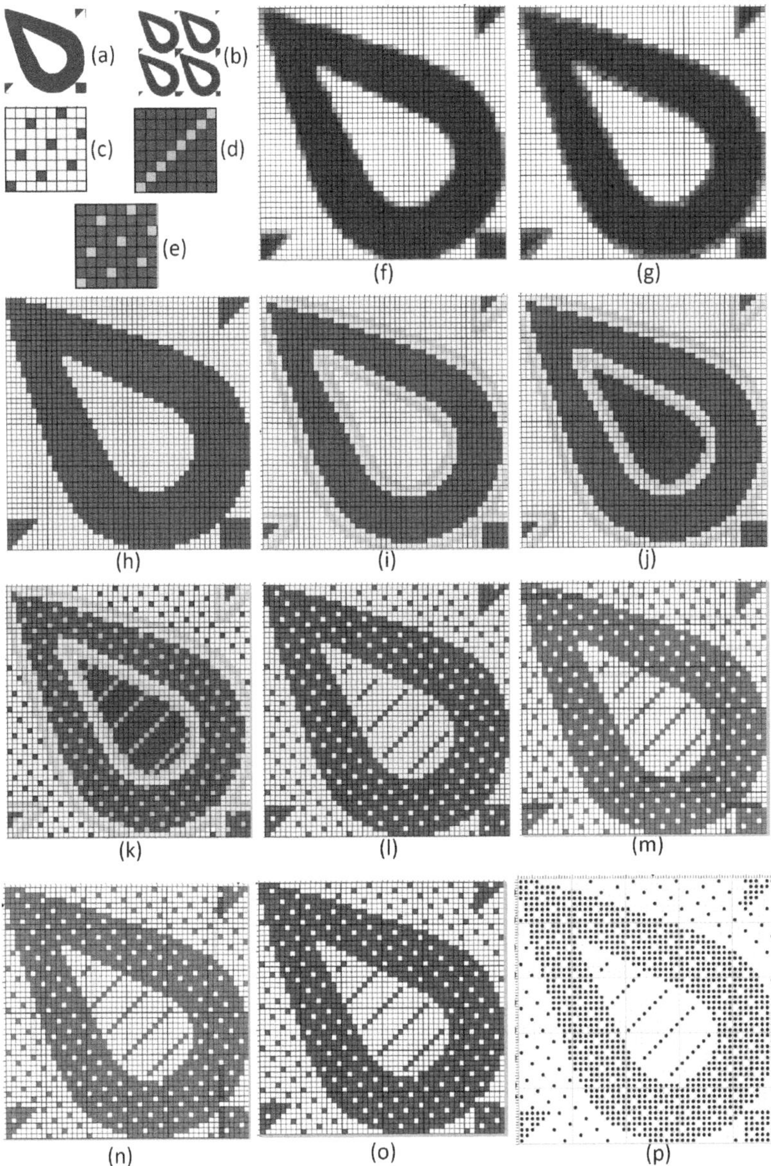

FIGURE 4.4 Different stages of computer-aided figured graph designing: (a) one repeat of the motif, (b) four repeats of the motif, (c) weave for the ground, (d) weave for the figure part, (e) weave for another ground part, (f) graph of the scanned motif, (g) editing the outline of the figure, (h) colouring the ground and figure parts, (i) contouring the ground and figure parts, (j) colouring the ground and figure parts as per weaves to apply, (k) mapping different weaves for different colours, (l) changing the weave marks and contours into warp and weft colours, (m) finding the lengthy horizontal floats and binding them, (n) finding the lengthy vertical floats and binding them, (o) final digital graph in warp and weft colour, and (p) graph printout from the digital graph.

(say, 48 ends × 48 picks). Then, the motif is created using drawing tools like lines, circles, rectangles, curves, etc., with the aid of a mouse or digitizer. (ii) The designer can create the new motif on paper using a pencil in the required size (say, 1-inch width × 1-inch length). This new motif is then scanned.

4.2.3   COPYING THE OLD MOTIF

Instead of drawing the new motif, the designer can copy the old motifs. There are two ways to copy the old motif: (i) The designer can select the suitable motif already available in the books. Then, it is taken for scanning. The motifs available on the website can also be used by copying them into the designing software. It can also be printed. Then, it is taken for scanning. (ii) The designer can also select a suitable motif from the figured cloth already woven or printed. Then, it is taken for scanning.

4.2.4   SCANNING THE MOTIF TO THE REQUIRED GRAPH SIZE

The scanner is the required input device for graph designers. The scanner copies the motif on the paper or cloth (a vector image), converts it into a pixel image (a raster image), and sends it to the computer. For example, when the designer scan the motif size of 1-inch × 1-inch (Figure 4.4a), keeping the scanner resolution at 48, the motif gets scanned into the pixel size of 48 × 48. When the designer zoom the image on the screen with the 'grid on' option, it becomes the graphical image on 48 ends × 48 picks, as shown in Figure 4.4f.

4.2.5   EDITING THE OUTLINES OF THE FIGURE

By scanning, the motif gets enlarged to the required ends and picks automatically. However, the scanned image is in many colours, according to the colour depth of the motif on the paper. It is seen in the graph shown in Figure 4.4f. Hence, it is necessary to outline the figure parts perfectly with a new colour, as per the shape of the figure parts. It is also essential to reduce the colours in the scanned image using colour-reducing tools. Drawing the border of the figure parts using curves, lines, circles, and drawing tools with the help of a digitizer pen or mouse is called 'Editing'. In the graph shown in Figure 4.4g, the figure outline is drawn perfectly in red pixels according to the shape.

4.2.6   CHECKING THE REPEAT JOINING

While the designer draws or edits a repeat of all-over designs, it is always necessary to check their joining. In any all-over design, the top-bottom and left-right sides must join properly. In the CAFGD software, there is an option called 'Repeat Mode Drawing' for checking the repeat joining. With this option, the designer can set four repeats of the image. It facilitates checking the proper joining of the repeat and correcting the improper joining, if any. After checking the repeat joining, the graph file is saved separately.

4.2.7 Colouring the Figure and Ground Parts

After editing the figure borders perfectly, all the figure parts and the borders are changed into one colour and the ground into another colour. Figure 4.4h shows the figure parts are changed to red and the ground parts to yellow. Red-colour figure parts show the warp-up, and yellow-colour ground parts show the weft-up (warp down). After completing the colouring, the graph file is saved separately. When this graph is printed as a motif, it is the 'colour simulation' of the edited graph.

4.2.8 Contouring/Bordering the Figure and Ground Parts

In the figured graph design for single cloth, the binding weave mark applied should control the long floats. To retain the perfection of the edited outline, the binding marks need not be given in the places where there are no longer floats. In CAFGD, the contouring/bordering option is used to avoid binding marks in non-required places. Before applying the weave, all the figure parts are given a boundary line with a different colour using the contouring option. Similarly, the ground parts are also given a boundary line with a different colour. In the graph shown in Figure 4.4i, the red figure parts are given contouring with a light blue colour, which also indicates warp-up. Likewise, the yellow ground parts are given contouring with the magenta colour, which also indicates weft-up. This contouring will enable the designer to apply the weave marks only on the figure and ground colours (red and yellow) and avoid applying any weave on the boundary colours (light blue and magenta). After completing the contouring, the graph file is saved separately.

4.2.9 Colouring the Figure and Ground Parts as per the Weaves

After contouring, the ground and figure parts inside the border lines are filled with different colours, as per the number of weaves decided to apply. It was decided to have: (i) an 8-thread sateen weave for the ground part outside the figure; (ii) a 1-up, 7-down twill weave for the ground part inside the figure; and (iii) an 8-thread satin weave for the figure. Hence, the ground part outside the figure is filled with one colour (yellow), the ground part inside the figure part is filled with another colour (brown), and the figure is filled with one more colour (red), as shown in the graph given in Figure 4.4j. The file is saved separately after completing the colouring.

4.2.10 Mapping Different Weaves for Different Colours

Weave mapping is the process of assigning/mapping/filling the required weaves for different figure and ground colour parts of the edited motif. For this, in the brush option, the weaves to be applied are selected from the weave library. The fill tool is used to fill the weave for the selected colour. By doing so, the colour gets replaced with the selected weave. Different weaves are applied one after the other for each colour part. As decided earlier, different weaves are applied for different colours, as detailed below:

1. The ground—outside the figure—in yellow colour is applied with the 8-thread sateen weave shown in Figure 4.4c.
2. The ground—inside the figure—in brown colour is mapped with the 1-up, 7-down twill weave shown in Figure 4.4d.
3. The figure filled with the red colour is mapped with the 8-thread satin weave shown in Figure 4.4e.

Figure 4.4k shows the graph filled with the weaves as above. The file is saved separately after completing the weave mapping.

4.2.11 CHANGING THE WEAVE MARKS INTO WARP AND WEFT COLOURS

The different weaves applied in the weave mapping process are in different colours. In all these weaves, all the colours that indicate end-up are converted into warp colours used in the fabric. Similarly, all the colours that indicate end-down (weft-up) are converted into weft colours used in the fabric. Assuming that the warp colour is red and the weft colour is yellow, the contouring and weave colours are changed to warp and weft colours, as detailed below:

1. The light blue colour used for contouring the figure (ends up) is changed to red.
2. In the figure part, the base red colour in the satin weave (ends up) is left as red.
3. In the ground part (outside the figure), the blue colour binding marks of the sateen weave (ends up) are changed to red.
4. In the ground part (inside the figure), the orange colour binding marks in the twill weave (ends up) are changed to red.
5. The magenta colour used for contouring the ground (ends down) is changed to yellow.
6. In the figure part, the green colour binding marks in the satin weave (ends down) are changed to yellow.
7. In the ground part (outside the figure), the base yellow colour in the sateen weave (ends down) is left as yellow.
8. In the ground part (inside the figure), the base brown colour in the twill weave (ends down) is changed to yellow.

The graph changed to only yellow and red colours as above is shown in Figure 4.4l. The file is saved separately after changing the colours.

4.2.12 IDENTIFYING THE LENGTHY FLOATS AND BINDING THEM

After weave mapping and re-colouring, it is necessary to find the lengthy warp and weft floats present, if any, and bind them suitably. The float-finding option is used to identify the floats. Then, the floats are given suitable binding marks using the weave fill tool. The graph shown in Figure 4.4m shows the lengthy horizontal floats identified in the ground and the figure and the floats are given binding as per the weave in which the float occurred. After binding the floats, once again, the colours are converted into yellow and red. The graph shown in Figure 4.4n shows the lengthy vertical floats identified in the ground, and the figure and the floats are given binding as per the weave in which the float occurred. After binding the floats, once again, the colours are converted into yellow and red. With this, the preparation of the figured graph with different weaves is completed with only two colours (red and yellow) as shown in Figure 4.4o. The graph file is saved separately. At this stage, if the graph is printed as a motif, it is called the 'colour and weave simulation' of the fabric.

4.2.13 Preparing and Storing the Digital Figured Graph in Black and White

The final work is to prepare the figured graph in black and white. For this, the warp colours are changed to black, and the weft colours are changed to white. It can also be vice versa. This digital figured graph in black and white is stored in a suitable file format.

4.2.14 Transferring the Digital Graph for Weaving, Punching, and Printing

The black-and-white digital graph is transferred directly to the electronic jacquard for weaving. Otherwise, it is transferred to an electronic card-punching machine to punch the cards for mechanical jacquard weaving. It is also used for graph printing. The graph is printed in a suitable graph count, with ends numbering, picks numbering, and converting the black colour into marks. The graph paper printout of the digital figured graph with circle marks for the end-up is shown in Figure 4.4p. This graph is taken to a manual hand-punching plate or a piano card-cutting machine to punch the cards for mechanical jacquard weaving.

4.3 COMPARISON BETWEEN THE MANUAL AND COMPUTER-AIDED FIGURED GRAPH DESIGNING

Table 4.1 shows the comparison between manual and CAFGD. The time involved in each stage of manual figured graph designing and CAFGD is highlighted to understand the advantages of using CAFGD.

The time taken for manual graph preparation and computer graphing may be almost the same for the smaller graph size, say 120 ends × 120 picks. When the graph size is larger, the time taken in manual graphing is longer, but the time taken in CAFGD is shorter. Changing the weave in a particular figure and ground part, storing the different stages of the graph, taking printouts of simulations at various stages of graph preparation, taking printouts of the final graph sheet, using it for manual punching, and transferring the digital file are the other advantages of CAFGD.

TABLE 4.1

Comparison between Manual and Computer-Aided Figured Graph Designing

Manual Figured Graph Designing	Computer-Aided Figured Graph Designing
The figured motif on the paper, and books are used for enlarging or the figured cloth already woven is also taken for enlarging.	The motif is drawn directly, or the motif on the paper, books, and motif downloaded from the website is used. The figured cloth already woven is also taken for scanning.
The motif is **divided** vertically and horizontally into a convenient number of segments for easy enlargement.	There is no need to divide the motif for enlargement.
The motif is **enlarged** on the graph of required ends and picks by drawing the curve following the motif outline. **It takes more time.**	The motif is **scanned** at the required resolution. The scanning directly enlarges the motif to the required ends and picks. Hence, **there is no need for enlargement. It takes less time**.
Enlarged curved lines are converted into **stepped-out lines** over the graph lines. **It takes a long time.**	Editing the scanned image is done by drawing the **pixel outline using another colour according to** the shape of the figure parts. **It is the only time-consuming process in computer graphing**.
A single colour is applied to the required figure parts. **It takes a long time**.	Different colours are applied in the figure and ground parts according to the number of weaves to apply **(one colour = one weave). It takes less time**.
Different **weaves** are applied to the figure and ground parts. **It takes a long time.**	One weave is assigned to one colour by taking the weave in the fill tool and clicking over the colour. **It takes less time**. This weave filling of CAFGD saves time when compared to manual graph designing.
While applying the weave marks, the designer avoids the marks in non-required places, as per the cloth variety.	The use of contouring or bordering options for the figure and ground parts avoids binding marks in the non-required places, as per the cloth variety.
Checking the repeat joining by rolling the graph sheet **takes a long time**.	Checking the repeat joining using repeat mode **takes less time**.
Different design arrangements like bi-symmetrical, multi-symmetrical, turn round, and dropping **takes a long time**.	Different design arrangements like bi-symmetrical, multi-symmetrical, turn round, and dropping **takes less time**.

BIBLIOGRAPHY

1. Ng, F., and Zhou, J. (2013). *Innovative Jacquard Textile Design Using Digital Technologies* (pp. 27–48). Woodhead Publishing.
2. Grosicki, Z. J. (2004). Construction of squared paper designs. In *Watson's Textile Design and Colour* (pp. 208–211). Woodhead Publishing Limited.
3. Mitra, A. (2014). CAD/CAM solutions for textile industry. *International Journal of Current Research and Academic Review*. 2(6), 41–50.
4. Panneerselvam, R. G. (2013). Use of MS paint for jacquard graph designing and printing. *Indian Journal of Fibre and Textile Research*. 38, 186–192.

5 Tools and Options of Figured Graph Editing Software

5.1 FIGURED GRAPH DESIGNING SOFTWARE AND EDITING SOFTWARE

Table 5.1 shows the list of a few Computer-Aided Figured Graph Designing (CAFGD) software. All the CAFGD software has two categories of tools and options. One is the general tools and options used for the preliminary colour editing and drawing of the graph, and the other is for advanced graph making for jacquard weaving. The colour editing and drawing tools, and options of all the CAFGD are based on the colour editing, drawing tools and options of common drawing and editing software like MS Paint, the default accessory of the MS Windows operating system, Paintshop Pro, and Adobe Photoshop. Hence, most of the CAFGD software uses one of these editing programs for basic colouring and editing of the graph. The CAFGD software is used only for advanced graph making. For example, Arahpaint and Arahweave are two modules of Arahne software for figured graph designing. Using Arahpaint, the designers can complete the graph editing in all aspects and the weave application. Arahweave is used for advanced weave applications and for creating other graph simulations. Moreover, Arahpaint is open-source software. Hence, the figured graph designers can use MS Paint and Arahpaint, the editing software, for basic colour editing, drawing, and the application of weaves on the figured graph.

The general tools in the figured graph editing software are grouped into drawing tools, selection tools, and graph area modifying and resizing tools. Figure 5.1 shows the icons of 32 tools arranged in serial order, with their serial numbers at the bottom from 1 to 32. The number is indicated as T1, T2 … and T32, where the

---

TABLE 5.1

List of a Few Computer-Aided Figured Graph Designing (CAFGD) Software

Name	Website	Origin	Name of the CAFGD Modules
Arahne	www.arahne.si	Slovenia	Arahpaint, Arahweave
Digibunai	www.digibunai.dic.gov.in/index.php/login-user	India	Artwork Designer
Nedgraphics	www.nedgraphics.com	Netherland	Jacquard CAD Solution
Pointcarre	www.pointcarre.in	France	Jacquard weaving software
Teckmen	www.teckmen.co.in	India	CADVantageWin Jacquard
Textronics	www.textronic.com	India	Design Jacquard

DOI: 10.1201/9781003441205-5

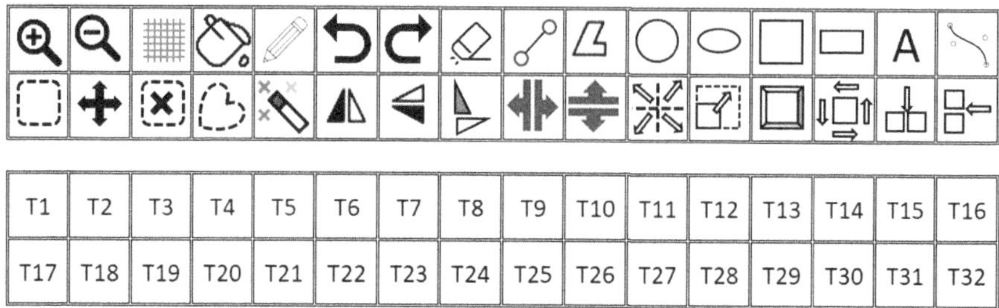

FIGURE 5.1   Icons of 32 general tools available in graph editing software.

letter 'T' indicates the word 'Tool'. Table 5.2 shows the serial number of the tools as per Figure 5.1, the corresponding name of the tools, and their use/function in graph editing. In the editing software, in general, most of the tools are used by starting the action with the left click of the mouse and dragging the mouse to do the actions like drawing and selecting. Right-clicking completes the action.

TABLE 5.2

Name and Use of 32 General Tools Available in Graph Editing Software Under Different Groups

S. No. of the Tools as per Figure 5.1	Name of the Tool	Use/Function
		I. Drawing Tools
T1	Zoom in	To increase the graph size on the monitor
T2	Zoom out	To decrease the graph size on the monitor
T3	Graph on and off	To open/close the graph lines over the white digital file
T4	Fill colour	To fill the selected graph pixel area with the selected colour
T5	Pencil, Freehand draw	To draw a free-form curve line in the required thickness
T6	Undo	To go back to the work carried out step by step
T7	Redo	To go forward to the work carried out step by step
T8	Eraser or Rubber	To change the pixel colour to the background colour
T9	Line, Draw Line	To draw lines at the required angles and thickness
T10	Polygon	To draw polygon shapes with multiple sides using connecting lines
T11	Circle	To draw the geometrical shape—Circle
T12	Ellipse	To draw the geometrical shape—Ellipse
T13	Square	To draw the geometrical shape—Square

(Continued)

TABLE 5.2 (*Continued*)

Name and Use of 32 General Tools Available in Graph Editing Software Under Different Groups

S. No. of the Tools as per Figure 5.1	Name of the Tool	Use/Function
T14	Rectangle	To draw the geometrical shape—Rectangle
T15	Text	To type the required text in the required font and size
T16	Bezier	To create curved lines in the required shape
II. Selection Tools		
T17	Rectangular selection	To select the required graph area in a rectangular or square shape for copying or moving
T18	Move	To move the selected graph area
T19	Deselect or Merge down	To deselect the selected graph area or to merge the layer into the background
T20	Freehand selection	To select the required graph area in any free-form shape for copying or moving
T21	Magic wand	To select a particular part of the graph drawn by connecting pixels of one colour, by clicking over any one pixel
III. Graph – Rotate, Reverse, Cut and Shift Tools		
T22	Mirror/Flip vertically or Reverse in end-way	To reverse the selected graph area vertically/end-way direction
T23	Mirror/Flip horizontally or Reverse in pick-way	To reverse the selected graph area horizontally/pick-way direction
T24	Rotate	To rotate the selected graph area in the required angle and direction
T25	Cut and shift in end-way	To cut the repeat vertically (end-way) in the required place, and shift the right part to the left
T26	Cut and shift in pick-way	To cut the repeat horizontally (pick-way) in the required place, and shift the top part to the bottom
T27	Cut and shift from centre	To cut the graph repeat in the centre and make it into four quarters. Then shift each quarter in the opposite direction
IV. Tool for Multiplying the Ends-Picks		
T28	Graph multiply	To multiply the ends and picks in the graph by doubling, trebling … required number of times
V. Tool for Increasing or Decreasing the Graph Size		
T29	Graph resize	To increase or decrease the graph size by adding or removing the ends and picks on the required boundary side of the graph area

(Continued)

TABLE 5.2 (*Continued*)

Name and Use of 32 General Tools Available in Graph Editing Software Under Different Groups

S. No. of the Tools as per Figure 5.1	Name of the Tool	Use/Function
	VI. Tool for Moving the Graph	
T30	Move graph	To move the graph area, end-way, that is, right to left or left to right. To move the graph area, pick-way, that is, top to bottom or bottom to top
	VII. Tools for Adding or Removing Ends-Picks Inside the Graph	
T31	Add or remove ends inside	To add or remove the ends in the inside area of the graph
T32	Add or remove picks inside	To add or remove the picks in the inside area of the graph

5.2 DRAWING TOOLS

Drawing tools are useful to create simple figured graphs directly on the monitor. For this, a new file is opened using the new file option with the required number of ends and picks. The file opens as a white sheet. The file size is enlarged using the 'zoom-in' tool (T1) and reduced using the 'zoom-out' tool (T2). After enlarging, the white sheet is made into a graph sheet, and vice versa, by the 'graph on and off' tool (T3). Figure 5.2a shows the 32 ends × 32 picks graph size opened as white pixels, zoomed in and made with the graph lines. It is the white colour when the default background colour is white. The white ground colour of the graph is changed to yellow using the 'fill colour' tool (T4), keeping yellow as foreground colour. A freehand line or curve is drawn with a 'pencil' tool (T5) using red colour. Figure 5.2b shows the several types of curves drawn. The top curve is drawn with unconnected single pixels (nose-touching pixels). The middle curve is drawn with connected single pixels (double pixels in the step-changing place) using the 'thick line' option. The bottom curve is drawn with connected double pixels (treble pixels in the step-changing place) using the 'thick line' option.

The 'undo' tool (T6) is used to see the actions carried out in the graph by going back step by step. The 'redo' tool (T7) is used after using the undo tool to restore all the actions that were previously undone using the undo tool. The 'eraser or rubber' tool (T8) is to change the colour of the pixel to the background colour, which is white by default. Hence, when used over the figure pixel in the foreground colour, it changes to the background colour (white).

Figure 5.2c shows the lines drawn at different angles using the 'line' tool (T9). Figure 5.2d shows the horizontal, vertical, and 45° angular lines drawn using the other option of the line tool. By using the 'polygon' tool (T10), connected lines are drawn to create different shapes. Figure 5.2e shows a star drawn using the polygon tool. A circle with a single-pixel outline is drawn by using the 'circle' tool (T11), and a filled circle is drawn by using the other option of the circle tool, as shown in Figure 5.2f. Similarly, an ellipse is drawn with a single-pixel outline by using the

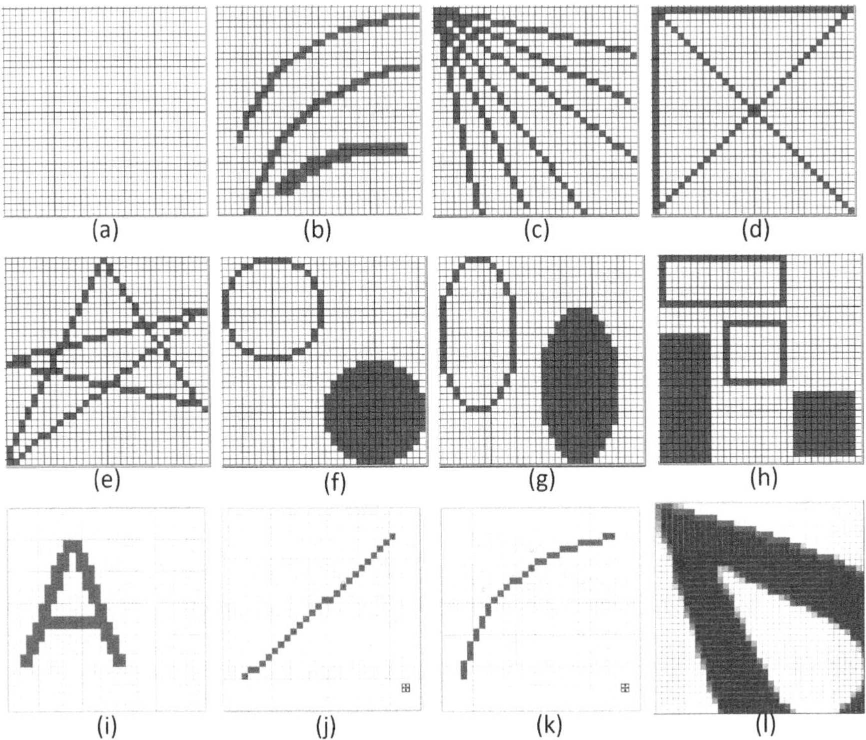

FIGURE 5.2 Graphs showing the use of drawing tools: (a) opening the graph file (32 × 32), (b) filling in the colour and drawing using the pencil tool, (c) drawing free lines with the line tool, (d) drawing angular lines with the line tool, (e) drawing star using the polygon tool, (f) drawing an outlined and filled circle with the circle tool, (g) drawing an outlined and filled ellipse with the ellipse tool, (h) drawing outlined and filled square and rectangle with the square and rectangle tool, (i) typing the text using the text tool, (j) drawing a line using the four-point bezier tool, (k) creating the curve from the bezier line, and (l) using the bezier line for editing the curves of the scanned image.

'ellipse' tool (T12), and a filled ellipse is drawn by using the other option of the ellipse tool, as shown in Figure 5.2g. Figure 5.2h shows the square and rectangle in a single-pixel outline drawn by using the 'square' tool (T13), and 'rectangle' tool (T14). The filled square and rectangle are drawn by using other options of these tools. Figure 5.2i shows the graphical form of the letter 'A' when typing the text 'A' using the 'text' tool (T15). Texts of different fonts and sizes can be typed using other options of text tools. Figure 5.2j–l shows the use of the 'bezier curve' tool (T16) to draw the curve. When the bezier tool is dragged between two points, a line is formed with four control points. The first point is at the start, two are in the middle, and the last one is at the end, as shown in Figure 5.2j. By moving the middle points in the required direction, the line changes to the required curve shape, as shown in Figure 5.2k. The bezier tool is used to draw the single-pixel, single-colour curved outline to edit the scanned image, as shown in Figure 5.2l.

5.3 SELECTION TOOLS

The selection tools are used to select a particular area of the graph or a selective figure part or multiple figure parts. When the new graph file is opened and the figure is drawn, it is taken as the fixed ground layer. By using the 'rectangular selection' tool (T17), when a part of the graph area is selected, the selected area becomes the movable second layer over the ground layer. The selected graph area includes both the figure and the ground. The selected second layer can be moved by using the 'move' tool (T18). While moving, both the red figure and yellow ground get moved, as seen in Figure 5.3a. While doing so, the ground yellow in the selected layer, hide the figure in the ground layer. The hiding of the red figure by the yellow ground is avoided by making the yellow colour transparent using the transparent option. Figure 5.3b shows the movement of the second layer after making the yellow colour transparent, where the red figure in the ground layer is fully visible. When the moving of the second layer is completed, the fixing and combining of two layers into one layer is done by the 'deselect or merge down' tool (T19). Figure 5.3c shows the selection of two adjacent circles by drawing a line around the two circles using the 'freehand selection' tool (T20). When the drawing is completed, both the circles get selected. They can be moved together to the required place. Figure 5.3d shows the selection of one of the two adjacent circles by the 'magic wand selection' tool (T21) for moving or copying. In the graph area, as many individual figure parts in similar colours as required can be selected using other options of the magic wand tool.

5.4 REVERSE, ROTATE, AND CUT-SHIFT TOOLS FOR GRAPH

Figure 5.4a shows a typical mango (Aambi/Aam) figured graph drawn on 32 × 32 with a circular red nose at the top left and a small green stem at the bottom right. By keeping this graph open and clicking the 'mirror/flip vertically or reverse in the end-way' tool (T22), the graph gets reversed in the end-way. That is, the first end to the last end of the graph in Figure 5.4a serially gets reversed into the last end to the first end, as seen in Figure 5.4b. By clicking the same tool again over this graph

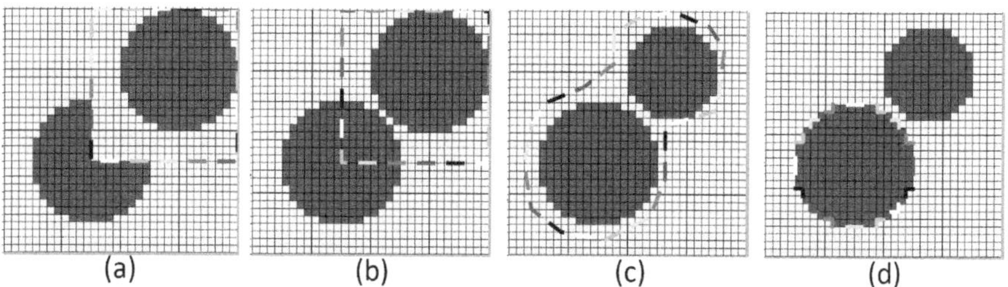

(a)   (b)   (c)   (d)

FIGURE 5.3 Graphs showing the use of selection **tools**: (a) selecting the figure and ground part using the rectangular selection tool, (b) moving only the selected figure part (without ground) using the move tool, (c) selecting the two adjacent figure parts using the free-form selection tool, and (d) selecting one of the two adjacent figure parts using the magic wand tool.

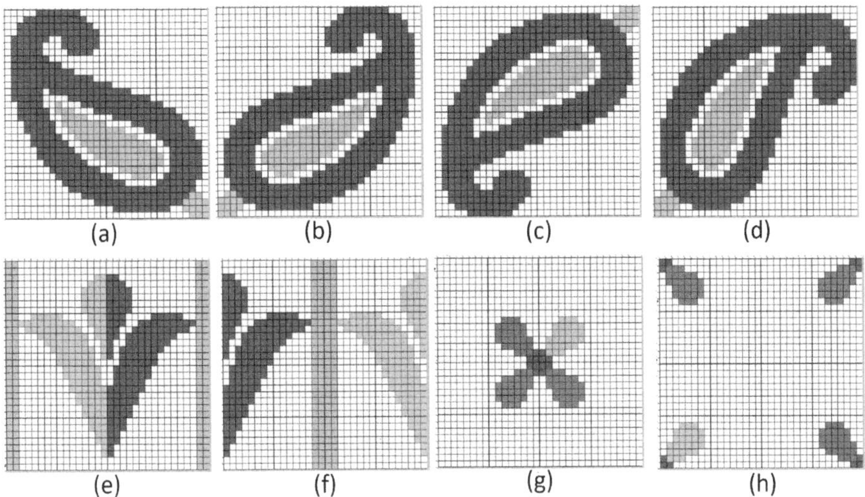

FIGURE 5.4 Graphs showing the use of reverse, rotate, and cut-shift tools for graph: (a) a figured graph on 32 × 32, (b) reversing Figure (a) in end-way, using the end-way reversing tool, (c) reversing Figure(a) in pick-way, using the pick-way reversing tool, (d) rotating the edited Figure(a) to 90° (clockwise), using the rotate tool, (e) a figured graph on 32 × 32 with the two ends blue border on the left and right side, (f) cutting in the centre (end-way) and shifting the right part to the left, using the cut and shift-end-way tool, (g) a figured graph drawn at the centre of 32 × 32, and (h) cutting the graph at the centre and shifting the figure to opposite corners, using the cut and shift from centre tool.

(Figure 5.4b), the graph gets reversed again, as seen in Figure 5.4a. By keeping the graph in Figure 5.4a open and clicking the 'mirror/flip horizontally or reverse in pick-way' tool (T23), the figure gets reversed in the pick-way. That is, the first pick to the last pick of the graph in Figure 5.4a serially gets reversed into the last pick to the first pick, as seen in Figure 5.4c. By clicking the same tool again over this graph (Figure 5.4c), the graph gets reversed back as seen in Figure 5.4a.

By keeping the graph in Figure 5.4a open and clicking the 'rotate' tool (T24), the figure gets rotated to 90° (clockwise direction). That is, the first pick to the last pick of the graph in Figure 5.4a, serially gets rotated from the first end to the last end, as seen in Figure 5.4d. By clicking the same tool again over this graph, the graph gets rotated again to 90° in the clockwise direction.

The 'cut and shift' tools (T25, T26) are useful for cutting the graph at the required place and shifting the two parts to opposite sides. By taking the 'cut and shift-end-way' tool (T25), and clicking in between any two ends of the graph, the graph is cut vertically into two parts at that place. Then, the left-side part of the graph gets shifted to the right, and the right-side part of the graph gets shifted to the left. In the same way, by taking the 'cut and shift-pick-way' tool (T26), and clicking in between any two picks of the graph, the graph is cut horizontally into two parts at that place. Then, the bottom side part of the graph gets shifted to the top side, and the top side part of the graph gets shifted to the bottom side. In Figure 5.4e, a figured graph on 32 ends × 32 picks is shown with a figure part in the centre and two ends bordered at the left and right sides in blue. By taking the 'cut and shift-end-way' tool (T25), and clicking at the centre of this graph, that is, in between the 16th and 17th ends, the graph is cut vertically. The left-side part gets shifted to the right, and the right-side part gets shifted to the left. By doing so, the right-side figure part (red) along with the blue line is shifted to the left, and the left-side figure part (green) along with the blue line is shifted to the right, as seen in Figure 5.4f. Similarly, the 'cut and shift-pick-way' tool (T26) can be used to cut and shift the graph in the pick-way.

The 'cut and shift from centre' tool (T27) does the work of both the cut and shift-end-way and cut and shift-pick-way tools together. By clicking this tool over the graph at any place, the graph gets cut vertically and horizontally in the centre, dividing the graph into four quarters. Furthermore, the bottom left part gets shifted to the top right, the top right part gets shifted to the bottom left, the bottom right part gets shifted to the top left, and the top left part gets shifted to the bottom right. Figure 5.4g shows a flower graph drawn in the centre with four petals in four assorted colours. By clicking over the graph with the 'cut and shift from centre' tool, the graph gets divided into four parts and shifted to opposite corners as explained above. The shifting is seen clearly in Figure 5.4h, where the four coloured petals of the flower get shifted to opposite corners.

5.5   TOOL FOR MULTIPLYING ENDS-PICKS

The 'graph multiply' tool (T28) is used to multiply the given graph size by doubling or trebling the ends or picks or both ends and picks. Otherwise, on reverse, the graph multiply tool is also used to decrease the size of the doubled or tripled ends or picks or both ends and picks of the graph into the graph of single-single ends or picks. A graph of 16 × 16 is shown in Figure 5.5a. By keeping this graph open and clicking the graph multiply tool, a dialogue box opens to input the required number of ends and picks. When 32 ends and 32 picks are entered as input, each end and pick of the graph in Figure 5.5a get doubled, resulting in the graph being 32 × 32 (16 × 2, 16 × 2), as shown in Figure 5.5b. When 16 ends and 32 picks are entered as input, each pick of the graph in Figure 5.5a gets doubled, leaving the ends unaltered, resulting in the

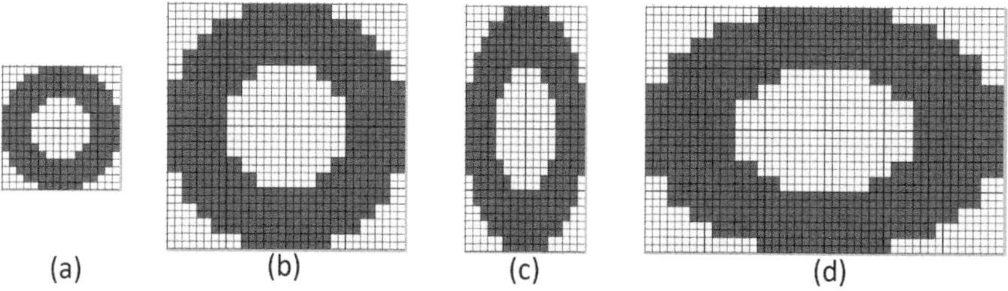

(a)   (b)   (c)   (d)

FIGURE 5.5   Graphs showing the use of the graph multiply tool for multiplying ends-picks (graph multiply). (a) a circle graph on 16 × 16, (b) graph (a) on 16 × 16 is made into 32 × 32 by doubling both ends and picks, (c), graph (a) on 16 × 16 is made into 16 × 32 by doubling only the picks, and (d) graph (a) on 16 × 16 is made into 48 × 32 by trebling the ends and doubling the picks.

graph being 16 ×32 (16 × 1, 16 × 2), as shown in Figure 5.5c. When 48 ends and 32 picks are entered as input, each end of the graph in Figure 5.5a gets tripled and each pick gets doubled, resulting in the graph being 48 × 32 (16 × 3, 16 × 2), as shown in Figure 5.5d.

On the reverse, by keeping the doubled ends and picks graph in Figure 5.5b (32 × 32) open, clicking the graph multiply tool, and entering the input as 16 ends and 16 picks, the ends and picks become single-single, resulting the graph to 16 × 16 (32/2, 32/2), as shown in Figure 5.5a. Again, keeping the doubled picks graph in Figure 5.5c (16 × 32) open, clicking the multiply tool, and entering the input as 16 ends and 16 picks, the picks become single-single, resulting in the graph to 16 × 16 (16, 32/2), as shown in Figure 5.5a. Similarly, keeping the trebled ends and doubled picks graph in Figure 5.5d (48 × 32) open, clicking the resize tool, and entering input as 16 ends and 16 picks, the ends and picks become single-single, resulting in the graph to 16 × 16 (48/3, 32/2), as shown in Figure 5.5a.

5.6   TOOL FOR INCREASING OR DECREASING THE GRAPH SIZE

The 'graph resize' tool (T29) is for increasing or decreasing the graph size by adding or removing the required number of ends/picks outside the graph area. While clicking this tool, a dialogue box opens to input the required number of ends and picks to be added or removed. The ends or picks get added or removed according to the graph size taken, the ends and picks input given, and the option/icon selected. The dialogue box has five options (icons), as shown in Figure 5.6a to add or remove ends/picks, as explained below:

- The topmost first icon (four square lines) is to add/remove ends and picks on all four sides (option-1).
- The second icon (right-top line) is to add/remove ends on the right side and picks on the top side (option-2).

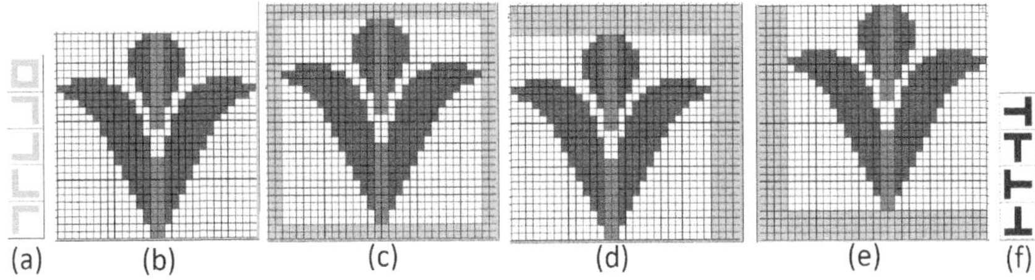

FIGURE 5.6   Graphs showing the use of graph resize tools and options for increasing or decreasing the graph size: (a) five options for adding/removing ends/picks, (b) a figure graph on 28 × 28 with double ends reversing in the centre (magenta colour), (c) adding two threads on all four sides of the graph (a) using add/remove threads outside (four sides) option-1, (d) adding four ends at the right and four picks on the top of the graph (a) using add/remove threads outside (right and top) option-2, (e) adding four ends on the left and four picks at the bottom of the graph (a) using add/remove threads outside (left and bottom) option-3, and (f) four options for moving ends/picks.

- The third icon (right-bottom line) is to add/remove ends on the right side and picks at the bottom side (option-3).
- The fourth icon (left-bottom line) is to add/remove ends on the left side and picks at the bottom side (option-4).
- The fifth icon (left-top line) is to add/remove ends on the left side and picks on the top side (option-5).

Table 5.3 shows the list of a few examples of adding/removing ends/picks outside the graph area using the 'graph resize' tool (T29) with reference to Figure 5.6b–e.

TABLE 5.3

List of a Few Examples Showing How to Add/Remove Ends/Picks Outside the Graph Area Using the Graph Resize Tool (T29) with Reference to Figure 5.6b–e

Option Number in Figure 5.6a	Original Size of the Graph Opened	Input - Final Size of the Graph (Original + or – Ends/Picks to Add/Remove)	Description of Adding and Removing Ends/Picks	Figure Reference
	28 × 28		28 × 28 graph is opened	Figure 5.6b
Option – 1 All sides	28 × 28	32 × 32 (28 + 2 + 2, 28 + 2 + 2)	2 ends are added on the right side	Figure 5.6c
			2 ends are added on the left side	
			2 picks are added on the top side	
			2 picks are added at the bottom side	
	28 × 28		28 ×28 graph is opened	Figure 5.6b
Option – 2 Right and Top sides	28 × 28	32 × 32 (28 + 4, 28 + 4)	4 ends are added on the right side	Figure 5.6d
			4 picks are added on the top side	
	28 × 28		28 ×28 graph is opened	Figure 5.6b
Option – 3 Right and bottom sides	28 ×28	36 × 40 (28 + 8, 28 + 12)	8 ends are added on the right side	
			12 picks are added at the bottom side	
	28 × 28		28 × 28 graph is opened	Figure 5.6b
Option – 4 Left and bottom sides	28 × 28	32 × 32 (28 + 4, 28 + 4)	4 ends are added on the left side	Figure 5.6e
			4 picks are added at the bottom side	
	28 × 28		28 ×28 graph is opened	Figure 5.6b

(Continued)

TABLE 5.3 (*Continued*)

List of a Few Examples Showing How to Add/Remove Ends/Picks Outside the Graph Area Using the Graph Resize Tool (T29) with Reference to Figure 5.6b–e

Option Number in Figure 5.6a	Original Size of the Graph Opened	Input - Final Size of the Graph (Original + or – Ends/Picks to Add/Remove)	Description of Adding and Removing Ends/Picks	Figure Reference
Option – 5 Left and top sides	28 × 28	36 × 40 (28 + 8, 28 + 12)	8 ends are added on the left side 12 picks are added on the top side	
	28 × 28		28 ×28 graph is opened	Figure 5.6b
Option – 2 Right and Top sides	28 × 28	28 × 36 (28 + 0, 28 + 8)	No ends are added on the right side Only 8 picks are added on the top side	
	28 × 28		28×28 graph is opened	Figure 5.6b
Option – 3 Right and bottom sides	28 × 28	40 × 28 (28 + 12, 28 + 0)	12 ends are added on the right side No picks are added at the bottom side	
	32 × 32		32 × 32 graph is opened	Figure 5.6c
Option – 1 All sides	32 × 32	28 × 28 (32 – 2 – 2, 32 – 2 – 2)	2 ends are removed on the right side 2 ends are removed on the left side 2 picks are removed on the top side 2 picks are removed at the bottom side	Figure 5.6b
	32 × 32		32 × 32 graph is opened	Figure 5.6d
Option – 2 Right and Top sides	32 × 32	28 × 28 (32 – 4, 32 – 4)	4 ends are removed on the right side 4 picks are removed on the top side	Figure 5.6b
	32 × 32		32 × 32 graph is opened	Figure 5.6e
Option – 4 Left and bottom sides	32 × 32	28 × 28 (32 – 4, 32 – 4)	4 ends are removed on the left side 4 picks are removed at the bottom side	Figure 5.6b

5.7 TOOL FOR MOVING THE GRAPH

By using the 'move graph' tool (T30), the graph can be moved to the required number of ends/picks in the desired direction. While clicking this tool, a dialogue box opens to input the required number of ends and/or picks. The ends or picks get moved according to the graph size taken, the ends and/or picks input given, and the option selected. The dialogue box also has four options, as shown in Figure 5.6f. The four icons are to move the graph without altering its size, as explained below:

- The topmost first option (the horizontal and right edge line) is to move the graph in the end direction towards the right.
- The second option (the vertical and top edge line) is to move the graph in the pick direction towards the top.
- The third option (the horizontal and left edge line) is to move the graph in the end direction towards the left.
- The fourth option (the vertical and bottom edge line) is to move the graph in the pick direction towards the bottom.

Table 5.4 shows the list of a few examples showing how to move the ends/picks of the graph in the required direction using the 'Move graph' tool (T30) with reference to Figure 5.6c–e.

TABLE 5.4

List of a Few Examples Showing How to Move Ends or Picks or Both Ends and Picks Using the 'Move Graph Area' Tool (T30) with Reference to Figure 5.6c–e

Option Number in Figure 5.6	Original / Final Size of the Graph	Input – Ends or Picks to Move	Description of Moving Ends/Picks	Figure Reference
	32 × 32		32 × 32 graph is opened	Figure 5.6d
Option – 1 horizontal right move	32 × 32	2 ends	The graph moves horizontally to the right side for 2 ends. Out of four blue ends on the right side, the last two blue ends (31 and 32) scroll and move to the left side as the first two ends (1 and 2), respectively, and	
Option – 2 Vertical top move	32 × 32	2 picks	the graph moves vertically to the top side for 2 picks. Out of four blue picks on the top side, the last two blue picks (31 and 32) scroll and move to the bottom side as the first two picks (1 and 2), respectively	Figure 5.6c

(Continued)

TABLE 5.4 (*Continued*)

List of a Few Examples Showing How to Move Ends or Picks or Both Ends and Picks Using the 'Move Graph Area' Tool (T30) with Reference to Figure 5.6c–e

Option Number in Figure 5.6	Original / Final Size of the Graph	Input – Ends or Picks to Move	Description of Moving Ends/Picks	Figure Reference
	32 × 32		32 × 32 graph is opened	Figure 5.6e
Option – 3 horizontal left move	32 × 32	2 ends	The graph moves horizontally to the left side for 2 ends. Out of four blue ends on the left side, the first two blue ends (1 and 2) scroll and move to the right side as the last two ends (31 and 32), respectively, and	
Option – 4 Vertical bottom move	32 × 32	2 picks	the graph moves vertically to the bottom side for 2 picks. Out of four blue picks at the bottom side, the first two blue picks (1 and 2) scroll and move to the top side as the last two picks (31 and 32), respectively	Figure 5.6c
	32 × 32		32 ×32 graph is opened	Figure 5.6c
Option – 3 horizontal left move	32 × 32	2 ends	The graph moves horizontally to the left side for 2 ends. The first two blue ends (1 and 2) scroll and move to the right side as the last two ends (31 and 32), respectively, and become 4 blue ends on the right side and	
Option – 4 Vertical bottom move	32 × 32	2 picks	the graph moves vertically to the bottom side for 2 picks. The first two blue picks (1, 2) scroll and move to the top side as the last two picks (31 and 32), respectively, and become 4 blue picks on the top side	Figure 5.6d
	32 × 32		32 × 32 graph is opened	Figure 5.6c
Option – 1 horizontal right move	32 × 32	2 ends	The graph moves horizontally to the right side for 2 ends. The last two blue ends (31, 32) scroll and move to the left side as the first two ends (1, 2), respectively, and become 4 blue ends on the left side and	

(*Continued*)

TABLE 5.4 (*Continued*)

List of a Few Examples Showing How to Move Ends or Picks or Both Ends and Picks Using the 'Move Graph Area' Tool (T30) with Reference to Figure 5.6c–e

Option Number in Figure 5.6	Original / Final Size of the Graph	Input – Ends or Picks to Move	Description of Moving Ends/Picks	Figure Reference
Option – 2 Vertical top move	32 × 32	2 picks	the graph moves vertically to the top side for 2 picks. The last two blue picks (31 and 32) scroll and move to the bottom side as the first two picks (1 and 2), respectively, and become 4 blue picks at the bottom side	Figure 5.6e

5.8 TOOLS FOR ADDING OR REMOVING THE ENDS-PICKS INSIDE THE GRAPH

The 'add or remove ends inside' (T31) and 'add or remove picks inside' (T32) tools are useful to add or remove the required number of ends or picks inside the graph area, before or after the given number of ends or picks. Figure 5.7a shows the dialogue box with different options open while clicking these tools. Since the ends are vertical, the vertical scissor with the (+) option is to add the desired number of ends, and the vertical scissor with the (−) option is to remove the desired number of ends. Since the picks are horizontal, the horizontal scissor with (+) option is to add the picks, and the horizontal scissor with (−) option is to remove the picks. The position of the ends or picks to be added or removed is entered in the 'position-spin button box' and the number of ends or picks to be added or removed is entered in

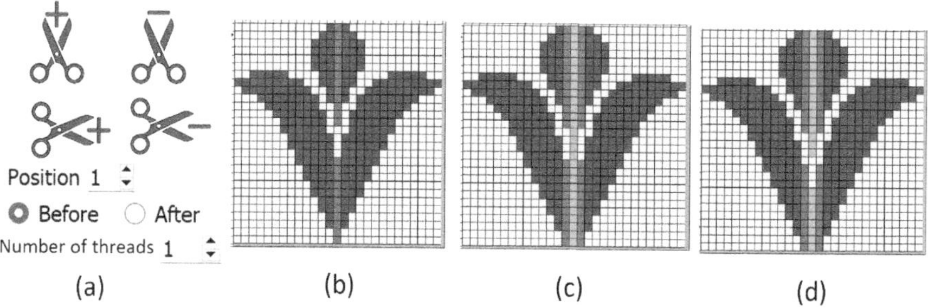

FIGURE 5.7 Graphs showing the use of tools and options to add/remove ends/picks inside the graph area: (a) a dialogue box with options to add/remove ends/picks inside the graph area, (b) removing one of the double ends in the centre reversing of the graph given in Figure 5.6b, (c) inserting an extra end in the middle of the double ends of the graph given in Figure 5.6b, and (d) correcting the interlacement of the centre end of the graph given in Figure 5.7c to have a single end reversing figured graph in the centre.

the 'number of threads-spin button box'. The adding or removing can be done either before or after the desired position of the ends or picks.

The 28 ends × 28 picks graph shown in Figure 5.6b is a bisymmetrical graph. Half of the left-side graph on 14 ends × 28 picks is reversed in the end direction and made into the full form of 28 × 28. While doing so, the reversing occurs with double ends, that is, with a similar interlacement of the 14th and 15th ends (shown in magenta colour). When this doubling in the centre is not desired, it is necessary to remove one of these two ends or add one more end in between these two ends.

If it is decided to remove the 14th end, it is to remove one end that is before the 15th end or after the 13th end. For this, the Add/remove ends inside tool (T31) is used. In the dialogue box, select the vertical scissor with the (−) option. Give '15' in the 'position-spin box' and select 'before radio button'. Give '1' in the 'number of threads-spin box'. By doing so, the 14th end gets removed, leaving only the 15th end, and the size of the graph becomes 27 ends × 28 picks, as shown in Figure 5.7b. Alternately, this can also be achieved by giving 13 in the 'position-spin box', selecting the 'after radio button' and giving 1 in the 'number of threads-spin box'. When it is decided to add one end in between the 14th and 15th end of the graph shown in Figure 5.6b, select the vertical scissor with the (+) option, give '14' in the 'position-spin box', select 'after radio button', and give '1' in the 'number of threads-spin box'. By doing so, one blank end gets added after the 14th end, as shown in green in Figure 5.7c, and the size of the graph becomes 29 ends ×28 picks. The newly added green end is edited as per the interlacement of the adjacent magenta ends, by which the graph becomes the single end reversing in the middle, as shown in Figure 5.7d.

For adding or removing picks, select the 'Add/remove picks inside' tool (T32), and the procedure followed to add or remove ends as described earlier is followed using the horizontal scissors.

5.9 COLOURING, DRAWING, LAYER, AND BRUSH TOOLS OPTIONS

Figure 5.8a shows the different options for the colouring tool. These options get activated when any drawing tool is selected for drawing. The top of Figure 5.8a shows: (i) Spin button box: used to change the thickness of the pixel while drawing with drawing tools. (ii) Out of the two colours shown, the top one is the foreground colour,

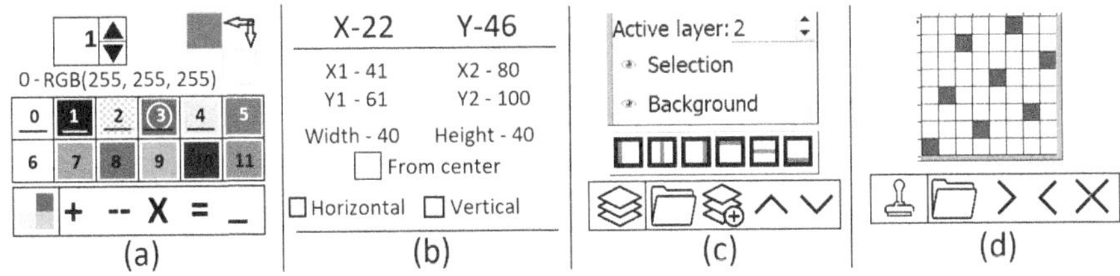

FIGURE 5.8 Different options for colouring, drawing, layer, and brush tools: (a) colouring tools options, (b) drawing tools options, (c) layer/selection tool options, and (d) brush tool options.

used with the left click of the mouse, and the bottom one is the background colour, used with the right click of the mouse. The colours can be interchanged by clicking the arrows connecting these two colour boxes. (iii) The figure shows '0', which is the number given for white, and its RGB value is 255, 255, 255. It indicates that the cursor or point of the tool is over the white colour pixel at a particular instant. (iv) In the middle is the display of the colour palette with its number. The first colour is white, and its number is zero. The second colour is black; its number is one, and so on. In the figure, the palette shows 12 colours with numbers from 0 to 11. (v) The red diamond lines on the colour number two (red) indicate that the red colour is made transparent. If the red colour is made transparent, when two layers are opened, the red colour in the top layer becomes invisible or transparent. By this, the designer could see the colours in the back layer below the red colour in the top layer. A particular colour is made transparent by clicking once over that colour. (vi) A circle mark seen over the colour number three (blue) shows that the blue colour is locked or protected. By doing so, the designer cannot change the blue colour used in the graph with other colours while drawing or filling. A particular colour is protected by double-clicking over that colour.

The bottom of Figure 5.8a shows icons for five options: add colour (+), remove the last colour in the palette (−), remove the unused colours (X), remove the duplicate colour in the palette (=), and show used colours (_). (vii) By clicking 'add colour' (+), a new colour gets added to the colour palette at the end of the colours already available. (viii) By clicking 'remove colour' (−), the last colour in the colour palette gets removed. (ix) By clicking 'remove duplicate colours' (=), the colours that repeat two or more times in the colour palette get removed, leaving behind the colour one time. (x) By clicking 'remove unused colours' (X), the colours not used in the active graph get removed, leaving only the used colours. (xi) By clicking 'show used colours' (_), the colours used in the active graph are indicated by an underline below the colours in the colour palette. In the figure, the underline shown below the colours from 0 to 4 indicates that these five colours are used in the active graph.

Figure 5.8b shows the different options for the drawing tool. These options get activated when any drawing tool is selected for drawing. (i) On the top of Figure 5.8b, X and Y indicate the end and pick number of the pixel at which the cursor or point of tool stays over the active graph. The figure shows X = 22 and Y = 46. It means that the cursor or point of the tool at a particular instant is at the 22nd end and 46th pick. (ii) At the bottom of Figure 5.8b, X1, Y1, and X2, Y2 indicate the starting and ending pixel numbers of the shape drawn. The size of the shape drawn is also shown in width pixels and length pixels. In the figure, X1 = 41 and Y1 = 61 indicate that the drawing of a shape (square) is started at this pixel number, and X2 = 80 and Y2 = 100 indicate that the drawing of a shape (square) is completed at this pixel number. At the bottom, the width becomes 40 and the length becomes 40 automatically. It shows that the shape (square) is drawn to the size of 40 ends (41–80) and 40 picks (61–100). (iii) By checking the 'From centre check box', different shapes like circles, ellipses, rectangles, and squares are drawn starting from the centre point of these shapes. (iv) By checking the 'Horizontal check box', if a line is drawn on only one end to the required length (picks), a horizontal rectangle is automatically obtained with the size equal to the length of the line drawn (picks) and the width equal to the full width (ends) of the graph size taken. (v) By checking the 'Vertical check box', if a line is drawn on only one pick to the required width (ends), a vertical rectangle is automatically obtained with the size equal to the width of the line drawn (ends) and the length equal to the full length (picks) of the graph size taken.

Figure 5.8c shows the different options of the layer or selection tool. These options get activated when the designer selects a graph part using the selection tool on the base graph. When selected, the selected area gets copied and becomes a layer, and the base graph becomes a background layer. Hence, (1) The active layer spin box shows two. (2) At the bottom, two 'eye icons' indicate that there are two layers. The selected layer is indicated as 'selection', and the background layer is indicated as 'background'. One of the layers can be made active by clicking over the 'eye icon' of a particular layer. (3) In the middle of the figure, six icons are given to move and paste the selected layer over the background layer in the required direction/position. The six icons are to move the selected layer to the following positions of the background layer: (i) the vertical red line on the left is used to move the layer to the extreme left of the ground layer, (ii) the vertical red line at the centre is to move the layer to the centre of the ground layer, (iii) the vertical red line on the right is to move the layer to the extreme right of the ground layer right, (iv) the horizontal red line on the top is to move the layer to the extreme top of the ground layer, (v) the horizontal red line in the middle is to move the layer to the middle of the ground layer, and (vi) the horizontal red line at the bottom is to move the layer to the extreme bottom of the ground layer. The bottom of Figure 5.8c shows icons for four options: load a layer from the file, load a new layer, move layers down, and bring layers up. (4) The 'Load layer from file' option is for opening the layer from the file over the ground layer. (5) The 'Load new layer' option is for opening a new layer with the required number of ends and picks over the ground layer. (6) When two layers are opened over the ground layer, the top layer is moved down to the other layer using the 'Move layer down' option, and the bottom layer is brought up over the other layer using the 'Bring layer up' option.

Figure 5.8d shows the different options for the brush tool. These options are used for selecting the weave graph as a layer, using it as a brush, and filling it fully on the other graph area using the fill tool. The top of Figure 5.8d shows the repeat of the weave opened. The bottom of Figure 5.8d shows the ways of opening the weave as a brush and the filling options that can be exercised while filling the weave on the graph. Load brush from file (folder icon), layer to brush (>), brush to layer (<), and delete brush (X) are the different options. (1) The 'Load brush from file' option is for loading the graph from the file as a brush. (2) The 'Layer to Brush' option is for transferring the selected graph layer from the active file graph to a brush graph. (3) The 'Brush to layer' option is for transferring the selected brush graph to the active file graph as a layer. (4) The 'Delete brush' option is for deleting the brush from the selected list.

5.10   LIST OF OPTIONS UNDER DIFFERENT MENUS

Table 5.5 shows the list of different options available under different menus in the editing software and their uses. The number is indicated as O1, O2 … and O47, where the letter 'O' indicates the word 'Option'.

TABLE 5.5

List of Different Options Available Under Different Menus in Editing Software and Their Uses

S. No.	Name of the Option	Use
		I. File Menu
O1	New picture…	To open a new graph in required ends and picks
O2	New window	To open a new window
O3	Load picture…	To load graph from general Windows Explorer
O4	Browse…	To browse selective graphs from the browse window and load it
O5	Load recent picture	To load the recent graph worked earlier
O6	Save picture	To save the graph in the existing name
O7	Save picture as…	To save the graph giving a new name
O8	Import vector image…	To import the pdf vector file as a raster file
O9	Load new layer…	To load a graph as a separate layer over the ground graph/layer
O10	Browse new layer…	To browse selective graphs from the browse window and load it as a new layer over the ground graph/layer
O11	Load recent layer	To load the recent layer graph worked earlier
O12	Save layer	To save the layer
O13	Exit program	To close and exit from the window
		II. Graph View Menu
O14	Grid Properties…	To set the thick and thin graph lines as per the count of the graph; To change the colour of the graph lines
O15	Show/Hide graph	To show the graph lines over the white file / To hide the graph lines over the white file
O16	Show/Hide thick lines	To show the thick graph lines over the thin lines as per the graph count / To hide the thick graph lines leaving the thin lines
		III. Edit Menu
O17	Undo	To go back to the work carried in step by step
O18	Redo	To bring forward the undo work carried in step by step
O19	Cut	To cut the selected graph area as design and keep it in clipboard memory
O20	Copy	To copy the selected graph area and keep it in the clipboard memory
O21	Paste	To paste the cut/copied graph area
O22	Duplicate	To copy and paste the selected graph area
O23	Select all	To select the entire graph area opened
O24	Invert selection	To select the area other than the selected area
O25	Clear	To apply the foreground colour in the selected area
O26	Delete layer	To delete the layer
		IV. Colour Menu
O27	Set number of colours…	To reduce the colours in the image/scanned image to the required number automatically

(Coninued)

TABLE 5.5 (*Continued*)

List of Different Options Available Under Different Menus in Editing Software and Their Uses

S. No.	Name of the Option	Use
O28	Reduce the number of colours...	To remove the non-required colours in the image by selecting the required colours or to reduce colours in black and white/Greyscale
O29	Colour remapping...	To remap/change one colour with another colour
O30	Sort colours	To sort the colours according to popularity, hue, lightness
O31	Protect colour	To protect the selected colour so that it could not be changed to other colours
O32	Transparent colour	To make the selected colour transparent so that the colour behind the selected colour is made visible
O33	Convert colours	To convert 256 colours to 16.7 million colours or vice versa
O34	Contour...	To apply inside/outside bordering to the figure and ground area
	V. Image Menu	
O35	Set in repeat mode...	To set multiple repeats with required reversing
O36	Duplicate X	To duplicate the selected graph area with the end-way reversing
O37	Duplicate Y	To duplicate the selected graph area with the pick-way reversing
O38	Multiply X	To multiply each end of the graph by the required number of times
O39	Multiply Y	To multiply each pick of the graph by the required number of times
O40	Smooth multiply	To multiply each end/pick of the graph to the required number of times and make the stepped outline smooth
O41	Shrink X	To shrink the graph size to 50% in the end-way
O42	Shrink Y	To shrink the graph size to 50% in the pick-way
O43	Smooth Shrink	To shrink the graph size to 50% in the end-way and pick-way and make the stepped outline smooth
O44	Halve X	To cut and remove 50% of the ends from the right side of the graph area
O45	Halve Y	To cut and remove 50% of the picks from the top side of the graph area
O46	Crop selection	To crop the selected graph area by removing the remaining area
O47	Trim	To remove the extra ground graph area on all four sides keeping only the ground area equal to the figure area
O48	Skew	To convert the quadrilateral shape of the graph area to the rectangle graph area

The icons and names of the tools and options shown in this chapter are general. It is not about any software. These may differ from software to software. When the designers use particular software, tools and options may be selected and used as per the purpose and functions performed as described in the examples.

BIBLIOGRAPHY

1. https://indiantextilejournal.com/software-used-in-textile-design/
2. https://windowsreport.com/free-software-jacquard-design/
3. Textronics Design Systems. "Design Jacquard: CAD/CAM for Jacquard Woven Fabrics." https://www.textronic.com/pdf/Design_Jacquard.pdf
4. Wonder Weaves Systems, "Wonder Weaves Systems: Products & Services," https://web.archive.org/web/20210317095745/ https://www.indiamart.com/wonder-weaves-systems-mumbai/services.html
5. Teckmen Systems, "Cad Vantage Win Jacquard," https://web.archive.org/web/20210317100135/https://cadvantagewin.com/jacquard.htm
6. Arahne, "Arahpaint for Windows," https://web.archive.org/web/20210317063213if_/ https://www.arahne.si/learn-support/ software-demo/arahpaint-for-windows/
7. Panneerselvam, R. G. (2013). Use of MS paint for jacquard graph designing and printing. *Indian Journal of Fibre and Textile Research*. 38, 186–192

6 Designing Different Types of Figured Graphs Using Various Tools and Options

6.1 DESIGNING FIGURED GRAPHS USING REVERSE AND ROTATE TOOLS

6.1.1 BI-SYMMETRICAL UNIT FIGURED GRAPH

Figure 6.1 shows the method of designing the end-way reversing bi-symmetrical unit figure and the pick-way reversing bi-symmetrical repeat figure using the reverse option by following the steps described below:

i. For designing the end-way bi-symmetrical figured graph, open the new graph file with the total ends and picks equal to the total ends and picks in the final graph. For example, if the final graph size required is 32 × 32, open the new graph file as 32 × 32.

ii. Draw half of the figure part to be reversed in the left area of 16 × 32; select this area using the rectangular selection tool (T17) and copy it as shown by dotted lines in Figure 6.1(a).

iii. Paste it as a layer; reverse the layer using the end-way reverse tool (T22).

iv. Make the ground colour transparent; move the layer to the right area of 16 × 32 and merge it with the ground, as shown in Figure 6.1b. This completes the designing of the end-way reversing bi-symmetrical unit figured graph.

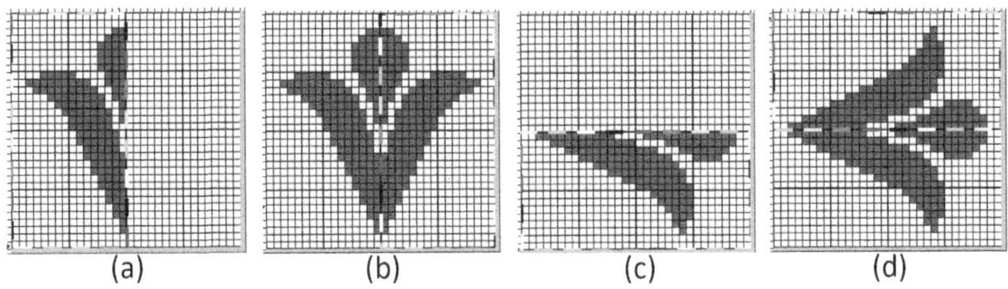

FIGURE 6.1 Designing the bi-symmetrical unit figured graph: (a) drawing half of the figure part on 16 × 32 on the left side and copying, (b) reversing the figure part in the end-way and merging it on the right side, (c) drawing half of the figure part on 32 × 16 at the bottom side and copying, and (d) reversing the figure part in the pick-way and merging it on the top side.

DOI: 10.1201/9781003441205-6

   v. For designing the pick-way bi-symmetrical figured graph, open the new graph file with the total ends and picks equal to the total ends and picks in the final graph. For example, if the final graph size required is 32 × 32, open the new graph file as 32 × 32.

  vi. Draw half of the total figure part in the bottom area of 32 × 16; select this area using the rectangular selection tool (T17) and copy it as shown by the dotted line in Figure 6.1c.

 vii. Paste the copied area as a layer; reverse the layer using the pick-way reverse tool (T23).

viii. Make the ground colour transparent; move the layer to the top area of 32 × 16 and merge it with the ground, as shown in Figure 6.1d. This completes the designing of the pick-way reversing bi-symmetrical unit figured graph.

Of course, the end-way reversing bi-symmetrical figure can be converted into a pick-way reversing bi-symmetrical figure by rotating it 90° or vice versa.

6.1.2   Multi-Symmetrical Unit Figured Graph

Figure 6.2 shows the method of designing the multi-symmetrical unit figure using the reverse option by following the steps described below:

    i. First, open the new graph file as per the repeat size required, say 32 × 32.

   ii. Draw 25% of the total figure part in the top-left area of 16 × 16; select and copy the top-left-side area as shown in Figure 6.2a.

  iii. Paste the copied area as a layer; reverse the layer using the end-way reverse tool (T22).

  iv. Make the ground colour transparent; move the layer to the top-right area of 16 × 16 and merge it with the ground, as shown in Figure 6.2b.

   v. Once again, paste the copied area as a layer; reverse the layer using the pick-way reverse tool (T23).

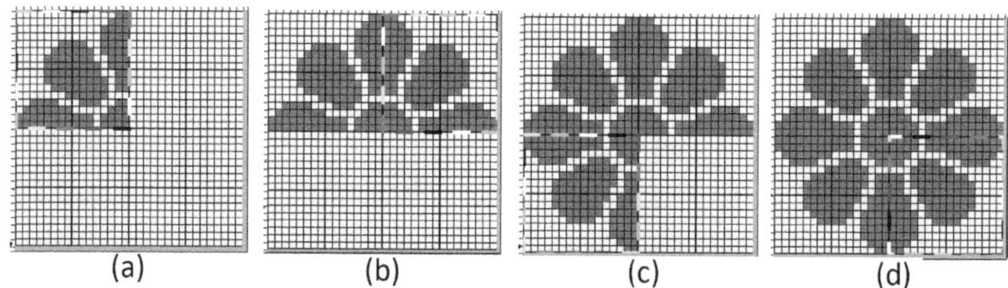

FIGURE 6.2 Designing the multi-symmetrical unit figured graph: (a) drawing 25% of the figure part on 16 × 16 on the top-left side and copying, (b) reversing the figure part in the end-way and merging it on the top-right side, (c) reversing the figure part in the pick-way and merging it at the bottom left side, and (d) reversing the figure part both in the end-way and pick-way and merging it at the bottom-right side.

    vi. Make the ground colour transparent; move the layer to the bottom-left area of 16 × 16 and merge it with the ground, as shown in Figure 6.2c.

   vii. Once again, paste the copied area as a layer; reverse the layer using the end-way reverse tool (T22); and again, reverse the layer using the pick-way reverse tool (T23).

  viii. Make the ground colour transparent; move the layer to the bottom-right area of 16 × 16 and merge it with the ground, as shown in Figure 6.2d. This completes the designing of the multi-symmetrical unit figured graph.

6.1.3 Turn-Around Unit Figured Graph

Figure 6.3 shows the method of designing the turn-around unit figured graph using the rotate option by following the steps described below.

    i. First, open the new graph file as per the repeat size required, say 32 × 32.

   ii. Draw the figure part that is to be turned around in the top-left area of 16 × 16; select and copy the top-left area as shown in Figure 6.3a.

  iii. Paste the copied area as a layer and turn it 90° by clicking the 90° rotating tool (T24) one time.

  iv. Make the ground colour transparent; move the layer to the top-right area and merge it with the ground, as shown in Figure 6.3b.

   v. Again, paste the copied area as a layer and turn it 180° by clicking the 90° rotating tool two times.

  vi. Make the ground colour transparent; move the layer to the bottom-right area and merge it with the ground, as shown in Figure 6.3c.

  vii. Once again, paste the copied area as a layer and turn it 270° by clicking the 90° rotating tool three times.

 viii. Make the ground colour transparent; move the layer to the bottom-left area and merge it with the ground, as shown in Figure 6.3d. This completes the designing of the turn-around unit figured graph.

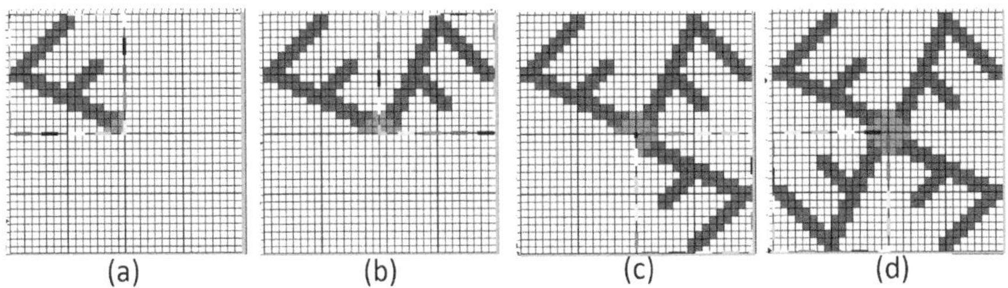

(a)   (b)   (c)   (d)

FIGURE 6.3 Designing the turn-round unit figured graph: (a) drawing 25% of the figure part on 16 × 16 on the top-left side and copying, (b) rotating the figure part 90° and merging it on the top-right side, (c) rotating the figure part 180° (90°, two times) and merging it at the bottom-right side, and (d) rotating the figure part 270° (90°, three times) and merging it at the bottom left side.

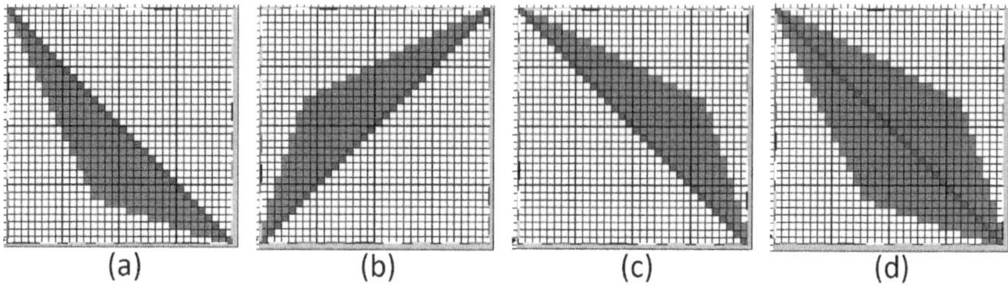

FIGURE 6.4   Designing the diagonally reversed unit figured graph: (a) drawing the figure part below the diagonal line and copying, (b) rotating the layer to 90°, (c) reversing the layer end-way, and (d) making the ground colour transparent and merging the layer by properly merging the diagonal line.

6.1.4   DIAGONAL REVERSE UNIT FIGURED GRAPH

Figure 6.4 shows the method of designing the diagonally reverse unit figured graph using the rotate and reverse options by following the steps explained below.

    i. First, open the new graph file as per the repeat size required, say 32 × 32.

    ii. Then, draw a right-to-left diagonal line (red).

    iii. Draw the figure below the red diagonal, as shown in Figure 6.4a.

    iv. Select and copy the file; paste it as a layer; turn the layer 90° round, as shown in Figure 6.4b.

    v. Then, reverse the layer in the end-way, as shown in Figure 6.4c.

    vi. Make the ground colour transparent; merge the layer over the ground. During merging, confirm the proper merging of the diagonal line in the layer with the diagonal in the ground, as shown in Figure 6.4d.

    vii. Finally, change the diagonal line colour as per the colour of the figure parts adjacent to it. With this, the designing of the diagonally reverse unit figured graph is completed.

6.2   DESIGNING FIGURED GRAPHS USING THE 'REPEAT MODE DRAWING BY REVERSING' OPTION

The 'Repeat Mode Drawing' option in the editing software is useful for drawing or editing the graph by arranging the repeats in the required type of reversing. In the repeat mode drawing by reversing, initially, a graph is opened with the required number of ends and picks, which represent the original repeat size. By the side of this original repeat, one or more duplicate repeats of the original repeat are opened in the end-way and pick-way. Then, these duplicate repeats are arranged in any one of the reversing modes, viz., straight, end-way bi-symmetrical, pick-way bi-symmetrical, or both end-way and pick-way multi-symmetrical. By keeping the duplicate repeats in the required order of arrangement, the designer can draw or paste the drawing in the first original repeat of the graph. Whatever the designer draws or pastes will automatically appear in the other duplicate repeats as per their order of arrangement.

When multiple repeats of the graph are arranged in the end-way, it is denoted as X1, X2, X3..., where 'X' represents the end-way repeating and the numbers 1, 2, 3... represent the serial number of repeats used. Similarly, when multiple repeats of the graph are arranged in the pick-way, it is denoted as Y1, Y2, Y3..., where 'Y' represents the pick-way repeating and the numbers 1, 2, 3... represent the serial number of repeats used.

After opening the original graph with the required number of ends and picks, when the designer selects the repeat mode option, a dialogue box opens, as shown in Figure 6.5a, with end-way repeat (X) − 1 and pick-way repeat (Y) − 1. It is taken as X1 and Y1. It represents one repeat. Hence, a single 'R' icon is opened in the open space, which is the original repeat with yellow shading. The designer can increase the number of repeats to be set in the end-way (X) and in the pick-way (Y). As per the number of repeats given, many 'Repeat icons' (R) get opened in the rectangular space provided. The required reversing of the repeats is set by reversing the icon. The reversing of the icon is done by clicking over the icon. For each click over the icon, the icon gets reversed in a different direction. For the first click, it is from straight to end-way reversing; for the second click, it is from end-way reversing to pick-way reversing; for the third click, it is from pick-way reversing to end-way + pick-way reversing; and for the fourth click, it is from end-way + pick-way reversing to straight.

When the input is X2, Y2—two repeats in the end-way and two repeats in the pick-way—that is, a total of four repeats (2 × 2) with four 'R icons' get opened, as shown in Figure 6.5b. The left-top 'R icon' is the original repeat, which is shown with yellow shading. The other three are duplicate repeats of the original repeat. Along with the original graph initially opened, three more duplicate graph repeats are opened, resulting in two repeats in the end-way and two repeats in the pick-way. All four 'R icons' pointing in one direction show that all four repeats are arranged straight. This repeat mode is used while drawing an 'all-over straight repeat' figured graph.

When the input is X2, Y1—two repeats in the end-way and one repeat in the pick-way—that is, a total of two repeats (2 × 1) with two 'R icons' get opened. The first 'R icon' is the original repeat, which is shown with yellow shading. The other one on the right side is the duplicate repeat. Along with the original graph initially opened, one more duplicate graph repeat is opened, resulting in two repeats in the end-way and one repeat in the pick-way. Initially, the two 'R icons' pointed in one direction. By clicking one time over the second R icon, it gets reversed in the end-way,

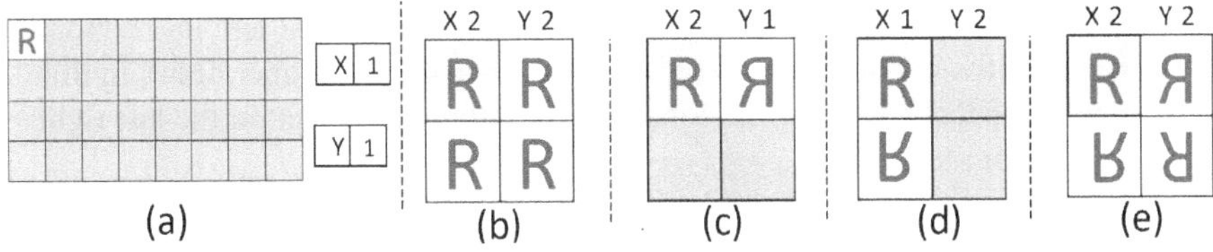

FIGURE 6.5 Repeat mode drawing by reversing option: (a) dialogue box for giving inputs, (b) repeat mode setting—straight, (c) repeat mode setting—end-way bi-symmetrical, (d) repeat mode setting—pick-way bi-symmetrical, and (e) repeat mode setting—end-way and pick-way multi-symmetrical.

as shown in Figure 6.5c. Now, the two graph repeats put together become end-way bi-symmetrical. This type of repeat mode setting is used for drawing the 'end-way bi-symmetrical' figured graph.

When the input is X1, Y2—one repeat in the end-way and two repeats in the pick-way—that is, a total of two repeats (1 × 2) with two 'R icons' get opened. The first 'R icon' is the original repeat, which is shown with yellow shading. The other one at the bottom is the duplicate repeat. Along with the original graph initially opened, one more duplicate graph repeat is opened, resulting in one repeat in the end way and two repeats in the pick way. Initially, the two 'R icons' point in one direction. By clicking two times over the second R icon, it gets reversed in the pick-way, as shown in Figure 6.5d. Now, the two graph repeats put together become pick-way bi-symmetrical. This type of repeat mode setting is used for drawing the 'pick-way bi-symmetrical' figured graph.

When the input is X2, Y2—two repeats in the end-way and two repeats in the pick-way—that is, a total of four repeats (2 × 2) with four 'R icons' get opened. The first 'R icon' is the original repeat, which is shown with yellow shading. The other three icons are duplicate repeats. Along with the original graph initially opened, three more duplicate graph repeats are opened, resulting in two repeats in the end-way and two repeats in the pick-way. Initially, the four 'R icons' point in one direction. By clicking one time over the top-right R icon, it gets reversed in the end-way; two times over the bottom-left R icon, it gets reversed pick-way; and three times over the bottom-right R icon, it gets reversed in both end-way and pick-way, as shown in Figure 6.5e. Now, the four graph repeats put together become multi-symmetrical. This type of repeat mode setting is used for drawing the 'multi-symmetrical' (end-way + pick-way) figured graph.

6.2.1 ALL-OVER STRAIGHT REPEAT-FIGURED GRAPH

Figure 6.6 shows the various stages of designing the all-over straight repeat-figured graph in repeat mode. The procedure for drawing the same is explained below.

   i. First, open the new graph file as per the required repeat size, say 48 × 48.
  ii. Open the repeat mode dialogue box and set the four repeats as straight repeats, as shown in Figure 6.5b. With this, the graph size increases to 96 × 96 and is divided into four repeats by the repeat lines. Figure 6.6a shows the graph divided into four repeats. The top-left area is the original first repeat. The other three repeats at the top right, bottom left, and bottom right are the duplicate graph repeat areas. When the designer draws in this repeat mode, he can see the continuity of the figure repeat on the left-right and bottom-top sides.
 iii. Draw the first vertical ellipse (red) at the centre of the top-left original repeat area. The ellipse appears automatically at the centre of all the other three duplicate repeats, as seen in Figure 6.6a.
  iv. Draw the second horizontal ellipse (green) on the right side of the first red ellipse. Draw it starting from the top-left repeat area and continuing to the top-right repeat area by crossing the vertical repeat line. While doing so, the

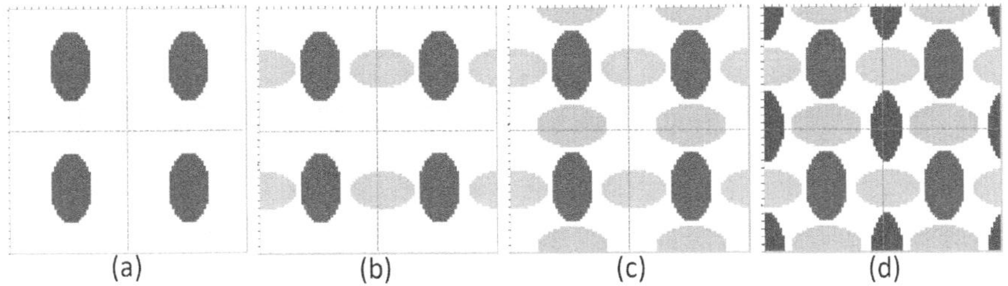

FIGURE 6.6 Designing the all-over straight repeat-figured graph in repeat mode: (a) drawing the first vertical orange ellipse at the centre of the first repeat, (b) drawing the second horizontal green ellipse starting from the top-left to top-right repeat crossing the vertical repeat line, (c) drawing the third horizontal orange ellipse starting from the top-left to bottom-left repeat crossing the horizontal repeat line, and (d) drawing the fourth blue ellipse starting from the top-left repeat to bottom right crossing both the vertical and horizontal repeat lines.

    green ellipse appears automatically in all the other repeat areas with proper continuity on the left-right side of all the repeats, as seen in Figure 6.6b.

v. Then, draw the third horizontal ellipse (orange) at the bottom side of the first red ellipse. Draw it starting from the top-left repeat area and continuing to the bottom-left repeat area by crossing the horizontal repeat line. While doing so, the orange ellipse appears automatically in all the other repeat areas with proper continuity on the top-bottom side of all the repeats, as seen in Figure 6.6c.

vi. Finally, draw the fourth vertical ellipse (blue) at the bottom of the second green ellipse. Draw it starting from the top-left repeat area. Continue it to the bottom-right area by crossing both the vertical and horizontal repeat lines at the centre. While doing so, the blue ellipse appears automatically in all the other repeat areas with proper continuity on the top-bottom and left-right sides of all the repeats, as seen in Figure 6.6d. This completes the designing of an all-over straight repeat-figured graph using the repeat mode with proper continuity on all four sides, left-right and bottom-top.

6.2.2 END-WAY BI-SYMMETRICAL REPEAT-FIGURED GRAPH

Figure 6.7 shows the different stages of designing the end-way bi-symmetrical repeat-figured graph in repeat mode. The procedure for drawing the same is explained below:

i. Decide the total ends and picks of the final end-way bi-symmetrical repeat-figured graph required. The total ends of the graph to open are equal to half of the total ends of the final graph. The total picks of the graph to open are equal to the total picks of the final graph. For example, if the final graph size is 48 × 48, the new graph file size opened is 24 ends (48/2) × 48 picks.

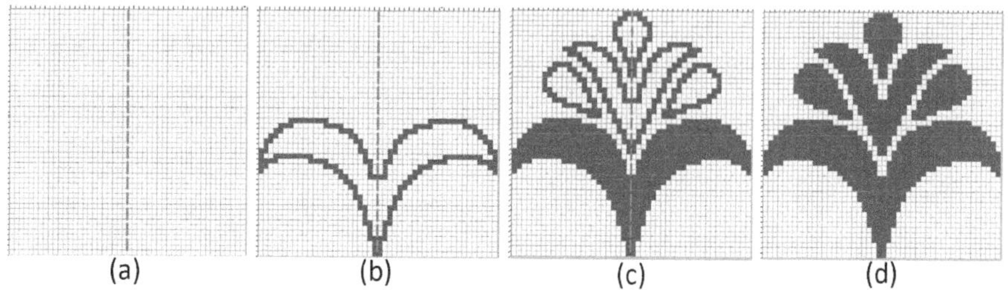

FIGURE 6.7   Designing the end-way bi-symmetrical repeat-figured graph in the repeat mode: (a) the graph file size increases to 48 × 48, and the repeat line divides the repeat into two vertical halves, (b) a figure outline drawn in the left-side area appears exactly reversed in the right-side area, (c) the right-side area gets filled automatically by filling the figure outlines in the left-side area, and (d) the end-way bi-symmetrical repeating graph is completed.

   ii. Open a new graph file of 24 ends × 48 picks.

  iii. Open the repeat mode dialogue box and set the two repeats as end-way bi-symmetrical repeats, as shown in Figure 6.5c. With this, the graph file size increases to 48 × 48. The repeat line divides the repeat into two vertical halves, as seen in Figure 6.7a. The left-side area is the original repeat. The right-side area is the duplicate, end-way reverse of the left-side area.

  iv. Keeping this repeat mode, draw the figure outline in the original left-side area. Simultaneously, the end-way reverse of the figure drawn will appear automatically in the right-side duplicate area. Figure 6.7b shows a figure outline drawn in the left-side area that appears exactly reversed in the right-side area.

   v. Draw the other figure outlines in the left-side area, as shown in Figure 6.7c.

  vi. Fill in the figure outlines in the left-side area. The figure outlines in the right-side area get filled automatically, as seen in Figure 6.7c.

 vii. Finally, make the left- and right-side areas into a single repeat. This completes the designing of the end-way bi-symmetrical repeat-figured graph, as shown in Figure 6.7d.

6.2.3   PICK-WAY BI-SYMMETRICAL REPEAT-FIGURED GRAPH

Figure 6.8 shows the different stages of designing the pick-way bi-symmetrical repeat-figured graph in the repeat mode. The procedure for drawing the same is explained below:

   i. Decide the total ends and picks of the final pick-way bi-symmetrical repeat-figured graph required. The total ends of the graph to open are equal to the total ends in the final graph. The total picks of the graph to open are equal to half of the total picks of the final graph. For example, if the final graph size is 48 × 48, the new graph file size opened is 48 ends × 24 (48/2) picks.

  ii. Open a new graph file of 48 ends × 24 picks.

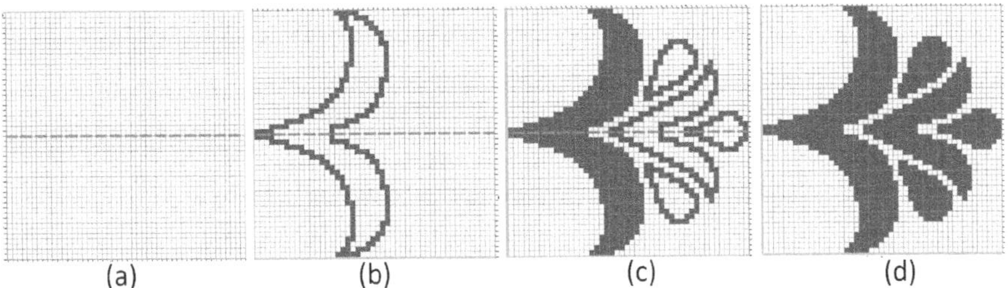

FIGURE 6.8 Designing the pick-way bi-symmetrical repeat-figured graph in the repeat mode: (a) the graph file size increases to 48 × 48, and the repeat line divides the repeat into two horizontal halves, (b) a figure outline drawn at the bottom side area appears exactly reverse in the top side area, (c) the top side area gets filled automatically by filling the figure outlines in the bottom side area, and (d) the pick-way bi-symmetrical repeating graph is completed.

iii. Open the repeat mode dialogue box and set the two repeats as pick-way bi-symmetrical repeats, as shown in Figure 6.5d. With this, the graph file size increases to 48 × 48. The repeat line divides the repeat into two horizontal halves, as seen in Figure 6.8a. The bottom side area is the original repeat. The top side area is the duplicate, pick-way reverse of the bottom side area.

iv. Keeping this repeat mode, draw the figure outline in the original bottom side area. Simultaneously, the pick-way reverse of the figure drawn will appear automatically in the top side duplicate area. Figure 6.8b shows a figure outline drawn in the bottom side area that appears exactly reversed in the top side area.

v. Draw the other figure outlines in the bottom side area, as shown in Figure 6.8c.

vi. Fill in the figure outlines in the bottom side area. The figure outlines in the top side area get filled automatically, as seen in Figure 6.8c.

vii. Finally, make the bottom and top side areas into a single repeat. This completes the designing of the pick-way bi-symmetrical repeat-figured graph, as shown in Figure 6.8d.It is important to understand that the procedures followed for drawing the end-way and pick-way bi-symmetrical repeat graphs are exactly opposite. Furthermore, it is also important to note that the pick-way bi-symmetrical repeat graph can be obtained by rotating the end-way bi-symmetrical repeat graph to 90°, or vice versa.

6.2.4 Multi-Symmetrical Repeat-Figured Graph

Figure 6.9 shows the method of designing the multi-symmetrical repeat-figured graph in repeat mode by following the steps described below.

i. Decide the total ends and picks of the required final multi-symmetrical repeat-figured graph. Open the new graph file with the ends and picks equal to half of the total ends and picks required in the final graph. For example,

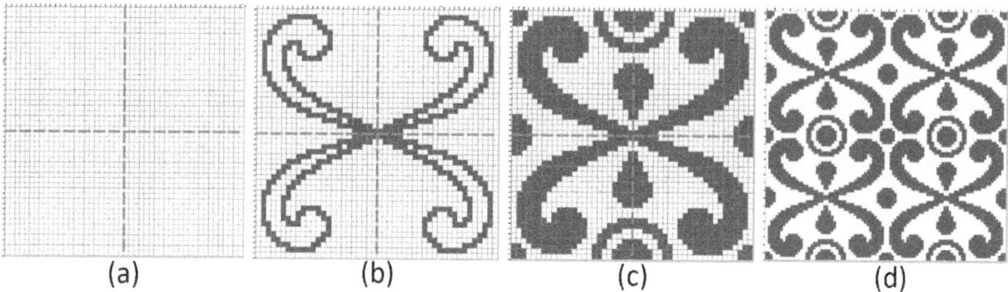

(a)   (b)   (c)   (d)

FIGURE 6.9 Designing the multi-symmetrical repeat-figured graph in the repeat mode: (a) the graph file size increases to 48 × 48 and gets divided into four parts by the vertical and horizontal repeat lines, (b) a figure outline drawn at the top-left area appears automatically in other repeats with the required reversing, (c) all the three repeat areas get filled automatically by filling the figure outlines in the top-left area, and (d) four repeats of the multi-symmetrical figured graph.

> if the final graph size is 48 × 48, the new graph file size opened is 24 (48/2) ends × 24 (48/2) picks.

  ii. Open a new graph file of 24 ends × 24 picks.

 iii. Open the repeat mode dialogue box and set the four repeats as multi-symmetrical repeats, as shown in Figure 6.5e. With this, the graph file size increases to 48 × 48 and gets divided into four repeats (24 × 24 each) by the vertical and horizontal repeat lines, as seen in Figure 6.9a. The top-left-side area is the original repeat. The other three are duplicates. The top-right-side area is the end-way reverse, the bottom-left-side area is the pick-way reverse, and the bottom-right-side area is the end-way and pick-way reverse of the top-left-side area.

  iv. Keeping this repeat mode, draw the figure outline in the original top-left-side area. Simultaneously, the end-way reverse of the figure drawn appears automatically in the top-right-side duplicate area. The pick-way reverse appears in the bottom-left-side area. The end-way and pick-way reverse appears in the bottom-right-side area, as seen in Figure 6.9b.

   v. Likewise, draw the rest of the figure outlines in the top-left-side area, as shown in Figure 6.9c.

  vi. Fill in the figure outlines in the top-left-side area. The figure outlines automatically get filled on the other three sides, as seen in Figure 6.9c.

 vii. Finally, make all four parts into a single repeat. This completes the designing of the multi-symmetrical figured graph.

viii. Figure 6.9d shows the four repeats of the multi-symmetrical figure.

It is important to understand that bi-symmetrical and multi-symmetrical figures can be designed using either the reverse tool or the repeat mode drawing option. The reverse tool is used when the design to be reversed has a perfectly edited outline and the shapes obtained after reversing are well defined. The repeat mode option is useful for the designers to draw the new shapes and know how they will look if they are reversed and joined in different directions. Hence, the designer should use the reverse tool and repeat mode drawing option as per the basic sketch and idea they have with them. Furthermore, when the designer draw the design in the original repeat, it appears in the duplicate repeats. It can also be vice-versa.

6.3 DESIGNING FIGURED GRAPHS USING THE 'REPEAT MODE DRAWING BY DROPPING' OPTION

The 'Repeat mode' option in the editing software is also useful to draw or edit the graph by dropping the repeats in the required fraction. In the repeat mode drawing by dropping, initially, a graph is opened with the required number of ends and picks, which represent the original repeat. By the side of this original repeat, one, two, or more duplicate repeats of the original repeat size are opened in the end-way or the pick-way. Then, these duplicate repeats are dropped/moved in any one of the fractions, like ½, ⅓, ¼, …. By keeping the duplicate repeats in the required fraction of dropping/moving, the designer has to draw or paste the drawing in the first original repeat of the graph. Whatever the designer draws or pastes will automatically appear in the other duplicate repeats as per their fraction of dropping/moving. Dropping the repeat is done with picks for the end-way repeats and moving the repeat is done with ends for the pick-way repeats.

For doing the pick-wise dropping, the number of repeats is increased in the end-way as per the fraction of dropping required by keeping one repeat in the pick-way. For one-half (½) of picks dropping, two repeats are used in the end-way and one repeat in the pick-way (X2, Y1). For one-third (⅓) of picks dropping, three repeats are used in the end-way and one repeat in the pick-way (X3, Y1), and so on. For doing the moving with ends, the number of repeats is increased in the pick-way as per the fraction of moving required by keeping one repeat in the end-way. For one-half (½) of ends moving, two repeats are used in the pick-way and one repeat in the end-way (X1, Y2). For one-third (⅓) of ends moving, three repeats are used in the pick-way and one repeat in the end-way (X1, Y3), and so on. Doing the dropping/moving both in the end-way and pick-way is called 'Half Drop'. For doing the half drop, the repeat size is increased both in the end-way and pick-way proportionately as per the figure size used.

6.3.1 PICK-WISE DROPPING FOR END-WAY REPEATING FIGURED GRAPH

Figure 6.10 shows the principles and graph designs of pick-wise dropping for end-way repeats. ½, ⅓, and ¼th dropping are explained by taking a small graph on 12 × 12. For the one-half (½) pick-wise drop, open the original graph, repeating on 12 × 12. Set the repeat mode dialogue box with two end-way repeats and one pick-way repeat (X2, Y1). Select the one-half drop (½) option. By doing so, the second repeat icon gets dropped down to half (50%) of the first repeat, as seen in Figure 6.10a. In the graph area, the second repeat (duplicate) gets opened in the end-way, with the repeat line dropping down for six picks (½ of 12) in the second repeat, as seen in Figure 6.10b. Draw the figure (ellipse) in the original repeat. It automatically gets dropped down for six picks and appears in the second repeat, as seen in Figure 6.10b. Then, make

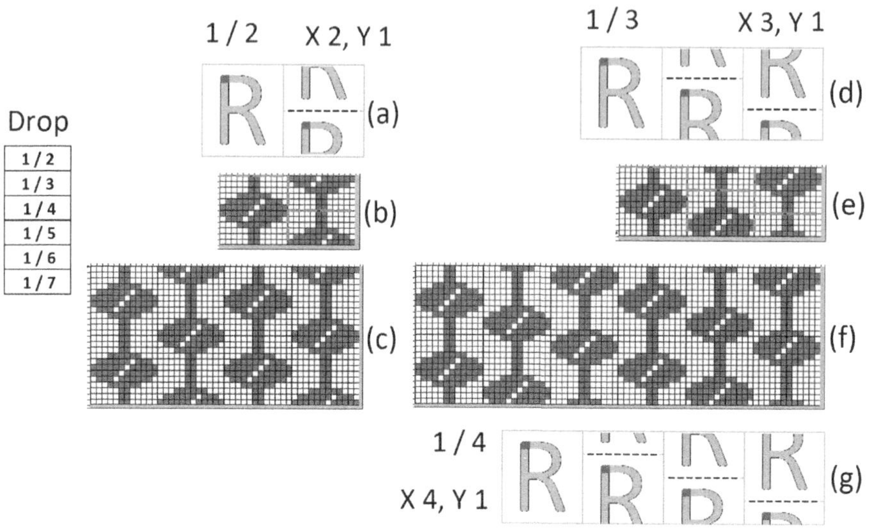

FIGURE 6.10 Repeat mode drawing—pick-wise dropping: (a) two repeats (two R icons) set in end-way with ½ of pick-wise dropping, (b) 12 × 12 graph design drawn in the first repeat automatically gets dropped to six picks in the second end-way repeat, (c) four repeats of ½ of pick-wise drop, (d) three repeats (three R icons) set in end-way with ⅓ of pick-wise dropping, (e) 12 × 12 graph design drawn in the first repeat automatically gets dropped to four picks each in the second and third end-way repeats, (f) four repeats of ⅓ of pick-wise drop, and (g) four repeats (four R icons) set in end-way with ¼ of pick-wise dropping.

these two repeats into a single repeat of 24 × 12, which is the one-half drop repeat. Figure 6.10c shows the four repeats on 48 × 24. From this graph, it is easy to understand how the ellipses are dropping down to the ½ fraction (six picks).

For the one-third pick-wise drop (⅓), open the original graph, repeating on 12 × 12. Set the repeat mode dialogue box with three end-way repeats and one pick-way repeat (X3, Y1). Select the ⅓ drop option. By doing so, the second repeat icon gets dropped to one-third (33.33%) of the first repeat, and the third repeat icon gets dropped to one-third (33.33%) of the second repeat, as seen in Figure 6.10d. In the graph area, the second and third repeats (duplicate) get opened in the end-way, and the repeat lines drop down for four picks (⅓ of 12) and eight picks (⅔ of 12), respectively, as seen in Figure 6.10e. Draw the figure (ellipse) in the first repeat. It automatically gets dropped to four picks in the second repeat and eight picks in the third repeat, as seen in Figure 6.10e. Then, make these three repeats into a single repeat of 36 × 12, which is the one-third drop repeat. Figure 6.10f shows the four repeats on 72 × 24. From this graph, it is easy to understand how the ellipses are pick-wise dropping down to one-third fraction (four picks each).

For the one-fourth pick-wise drop, open the original graph, repeating on 12 × 12. Set the repeat mode dialogue box with four end-way repeats and one pick-way repeat (X4, Y1). Select the one-fourth drop option. By doing so, the second repeat icon gets dropped down to one-fourth (25%) of the first repeat, the third repeat icon gets dropped down to one-fourth (25%) of the second repeat, and the fourth repeat icon gets dropped down to one-fourth (25%) of the third repeat, as seen in Figure 6.10g.

6.3.2 END-WISE MOVING FOR PICK-WAY REPEATING FIGURED GRAPH

Figure 6.11 shows the principles and graph designs of end-wise moving for pick-way repeats. ½, ⅓, and ¼ᵗʰ moving are explained by taking a small graph repeating on 12 × 12. For the ½ end-wise move, open the original graph, repeating on 12 × 12. Set the repeat mode dialogue box with two pick-way repeats and one end-way repeat (X1, Y2). Select the one-half drop (½) end-wise option. By doing so, the second repeat icon gets moved to half (50%) of the first repeat, as seen in Figure 6.11a. In the graph area, the second repeat (duplicate) gets opened in the pick-way, with the repeat line moving right for six ends (½ of 12) in the second repeat, as seen in Figure 6.11b. Draw the figure (ellipse) in the original repeat. It automatically gets moved right to six ends and appears in the second repeat, as seen in Figure 6.11b. Then, make these two repeats into a single repeat of 12 × 24, which is the one-half drop repeat. Figure 6.11c shows the four repeats on 24 × 48. From this graph, it is easy to understand how the ellipses are moving right to a one-half fraction (six ends each).

For the one-third end-wise move, open the original graph, repeating on 12 × 12. Set the repeat mode dialogue box with three pick-way repeats and one end-way repeat (X1, Y3). Select the one-third end-wise move option. By doing so, the second repeat

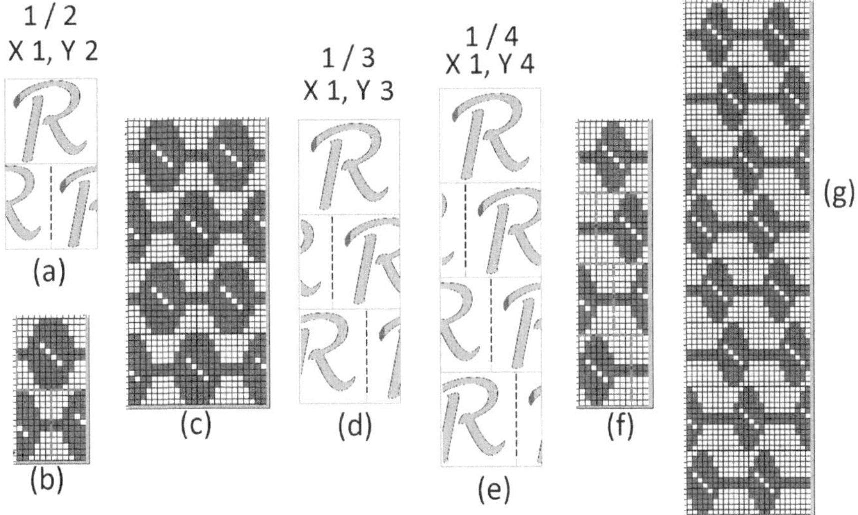

FIGURE 6.11 Repeat mode drawing—end-wise moving: (a) two repeats (two R icons) set in pick-way with ½ of end-wise moving, (b) 12 × 12 graph design drawn in the first repeat automatically gets moved right to six ends in the second repeat, (c) four repeats of ½ of end-wise move, (d) three repeats (three R icons) set in pick-way with ⅓ of end-wise moving, (e) four repeats (four R icons) set in pick-way with ¼ of end-wise moving, (f) 12 × 12 graph design drawn in the first repeat automatically gets moved to three ends in the second, third, and fourth end-way repeats, and (g) four repeats of ¼ of end-wise move.

icon gets moved right to one-third (33.33%) of the first repeat, and the third repeat icon gets moved right to one-third (33.33%) of the second repeat, as seen in Figure 6.11d.

For the one-fourth end-wise move, open the original graph, repeating on 12 × 12. Set the repeat mode dialogue box with four pick-way repeats and one end-way repeat (X1, Y4). Select the one-fourth end-wise move option. By doing so, the second repeat icon gets moved right to one-fourth (25%) of the first repeat, the third repeat icon gets moved right to one-fourth (25%) of the second repeat, and the fourth repeat icon gets moved to one-fourth (25%) of the third repeat, as seen in Figure 6.11e. In the graph area, the second, third, and fourth repeats (duplicate) get opened in the pick-way, with the repeat line moving right after three ends each (¼ of 12), as seen in Figure 6.11f. Draw the figure (ellipse) in the original repeat. It automatically gets moved to three ends each (¼ of 12) on the right side in the second repeat, six ends (²⁄₄ of 12) in the third repeat, and nine ends (¾ of 12) in the fourth repeat, as seen in Figure 6.11f. Then, make these four repeats into a single repeat of 12 × 48, which is the one-fourth drop repeat. Figure 6.11g shows the four repeats on 24 × 96. From this graph, it is easy to understand how the ellipses are moving to the one-fourth fraction (three ends).

As a common rule, the number of repeats opened for the drop/move is equal to the dropping fraction. If it is a one-half fraction, two repeats are opened; if it is a one-third fraction, three repeats are opened, and so on. The number of threads in the repeat must be a multiple of the fraction selected for dropping/moving. For the one-half fraction, the number of threads in the repeat must be a multiple of 2, that is, 8, 10, 12,... 120. For a one-third fraction, the number of threads in the repeat must be a multiple of 3, that is, 9, 12, 15,... 120. For the one-eighth fraction, the number of threads in the repeat must be a multiple of 8, that is, 8, 16, 24,... 120.

6.4 DESIGNING THE FIGURED GRAPH IN HALF-DROP PRINCIPLE

The editing software has the option to design the figured graph using the half-drop principle. It is possible to design two types of half-drops, namely half-drop straight and half-drop reverse. The procedures followed in designing these two types of figured graphs are given below.

6.4.1 Half-Drop Straight Principle

Figure 6.12 shows the method of designing the figured graph in the half-drop straight principle using the 'cut and shift from centre' option by following the steps described below.

    i. First, open the new graph file as per the repeat size required, say 32 × 32.
   ii. Draw a figure in the centre of the graph area as shown in Figure 6.12a. For example, an abstract bird with its nose towards the right side is drawn in the centre of the graph area. The figure is coloured with four colours such that it gets divided into four parts from its centre, viz. (i) right-top (red), (ii) right-bottom (green), (iii) left-bottom (purple), and (iv) left-top (blue). This colouring is for understanding the principle of half drop easily.
  iii. Select the graph area using the 'select all' option (ctrl + A); a copy of the graph is opened as a layer on top of the original graph (the ground).

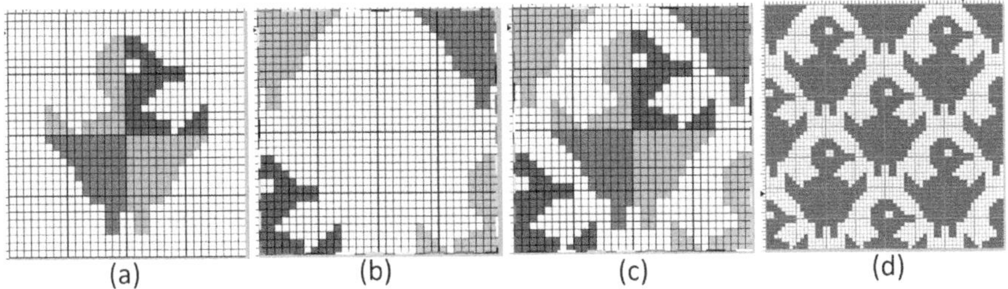

FIGURE 6.12 Designing the figured graph in the half-drop straight principle: (a) drawing the figure part in the centre of the graph area, selecting and making two layers, (b) the figure in the centre of the top layer cut into four parts and shifted from the centre to four corners of the repeat, (c) the figure in the centre of the ground layer becomes visible along with the four-figure parts in the corners of the top layer, by making the ground colour transparent, and (d) four repeats of the half-drop straight confirm the joining of the second corner unit figure.

    iv. Click on the 'cut and shift from the centre' tool (T27). By clicking, the figure in the centre of the top layer gets cut into four parts and shifted from the centre to the four corners of the repeat, as seen in Figure 6.12b. The (i) right-top part of the figure (red) gets shifted to the left-bottom corner of the repeat, (ii) the right-bottom part of the figure (green) gets shifted to the left-top corner of the repeat, (iii) the left-bottom part of the figure (purple) gets shifted to the right-top corner of the repeat, and (iv) the left-top part of the figure (blue) gets shifted to the right-bottom corner of the repeat.

    v. Make the ground colour of the top layer transparent. By doing this, the figure in the centre of the ground layer becomes visible along with the four-figure parts in the corners of the top layer, as seen in Figure 6.12c. Merge both layers into one. This is the final repeat of the figured graph in a half-drop straight principle with two same figure units in one repeat, one at the centre and another in four parts placed straight, similar to the centre, in the four corners of the repeat.

    vi. Figure 6.12d shows the four repeats of the half-drop straight. It is a half-drop straight repeat because the first unit and the second unit in the repeat are in the same direction. That is, both the birds are with their noses towards the right side. The four repeats also confirm the proper joining of the second corner unit figure and show the figure units are straight, that is, in one direction.

6.4.2 HALF-DROP REVERSE PRINCIPLE

Figure 6.13 shows the method of designing the figured graph in half-drop reverse principle using the 'cut and shift from centre' option by following the steps described below.

    i. First, open the new graph file as per the repeat size required, say 32 × 32.

    ii. Draw a figure in the centre of the graph area, as shown in Figure 6.13a. For example, an abstract bird with its nose towards the right side is drawn in the centre of the graph area. The figure is coloured with four colours such that it gets divided into four parts from its centre, viz. (i) right-top (red),

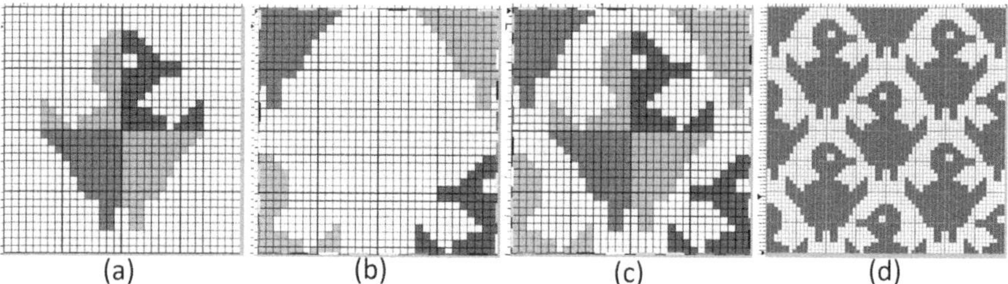

FIGURE 6.13 Designing the figured graph in the half-drop reverse principle: (a) drawing the figure part in the centre of the graph area, selecting, and making two layers, (b) the figure in the centre of the top layer cut into four parts, shifted from centre to four corners of the repeat and reversed in end-way, (c) the figure in the centre of the ground layer becomes visible along with the four-figure parts in the corners of the top layer by making the ground colour transparent, and (d) four repeats of the half-drop reverse confirm the proper joining of the second corner unit figure.

      (ii) right-bottom (green), (iii) left-bottom (purple), and (iv) left-top (blue). This colouring is for understanding the principle of half drop easily.

iii. Select the graph area using the 'select all' option (Ctrl + A); a copy of the graph is opened as a layer on top of the original graph (the ground).

iv. Click on the 'cut and shift from centre' tool (T27). By clicking, the figure in the centre of the top layer gets cut into four parts and shifted from the centre to the four corners of the repeat. *Reverse the layer using the end-way reverse tool.* By doing so, the four-figure parts in the corners get reversed, as seen in Figure 6.13b. The (i) right-top part of the figure (red) gets shifted and reversed to the right-bottom corner of the repeat, (ii) the right-bottom part of the figure (green) gets shifted and reversed to the right-top corner of the repeat, (iii) the left-bottom part of the figure (purple) gets shifted and reversed to the left-top corner of the repeat, and (iv) the left-top part of the figure (blue) gets shifted and reversed to the left-bottom corner of the repeat.

v. Make the ground colour of the top layer transparent. By doing this, the figure in the centre of the ground layer becomes visible along with the four-figure parts in the corners of the top layer, as seen in Figure 6.13c. Merge both layers into one. This is the final repeat of the figured graph in half-drop reverse principle, having two figure units in one repeat, one at the centre and another in four parts placed in the reverse direction in the four corners of the repeat.

vi. Figure 6.13d shows the four repeats of the half-drop reverse. It is a half-drop reverse because the first unit and the second unit are not in the same direction. That is, the first bird is with its nose towards the right side, whereas the second bird is reversed, and hence, it is with its nose towards the left side. The four repeats also confirm the proper joining of the second corner unit figure and show the figure units reversed in the opposite alternately in the diagonal direction.

The steps followed in designing half-drop straight and half-drop reverse-figured graphs are the same except for the third step. In half-drop straight, in the third step, after using the 'cut and shift from centre' tool (T27), the fourth step is followed. Whereas, in half-drop reverse, in the third step, after using the 'cut and shift from centre' tool (T27), the layer is reversed using the end-way reverse tool, which makes the second unit to get reversed. Then follows the fourth step.

Other types of half drop can be designed by doing the pick-way reverse or rotating instead of end-way reversing. This makes the second unit get reversed in the pick-way or rotated in the required angles.

BIBLIOGRAPHY

1. Ng, F., and Zhou, J. (2013). *Innovative Jacquard Textile Design Using Digital Technologies*. Elsevier Science. ISBN: 9780857098702, 0857098705.
2. Grosicki, Z. J. (2004). Construction and development of jacquard designs. In *Watson's Textile Design and Colour* (pp. 231–248). Woodhead Publishing Limited.
3. Grosicki, Z. J. (2004). Arrangement of figures. In *Watson's Textile Design and Colour* (pp. 249–293). Woodhead Publishing Limited.
4. Holyoke, J. (2013). *Digital Jacquard Design*. Bloomsbury Academic. ISBN: 9780857853455, 0857853457.

7 Scanning, Resizing, Colour Reduction, Editing, and Weave Mapping

7.1 SCANNING THE FIGURED MOTIF

The scanner is a digital input device for graph designers. It scans the vector-figured image/motif on the paper or cloth and converts it into a raster image in the required pixelated graphical format. The scanner is equipped with a glass top and an electronic scanning carriage to read the pictures kept on the glass top, as shown in Figure 7.1a. It has a cover that must be lifted to place the pictures to be scanned. There is a reflective mat underside of the top cover of the scanner. A typical scanner is shown in Figure 7.1a, viable for scanning the figured motif on paper, books, and cloth, that are bulky. Generally, the scanner comprises two components: the visible one is the hardware, while the other invisible component is the scanner software. This scanner is connected to the computer, and the scanner software is installed on

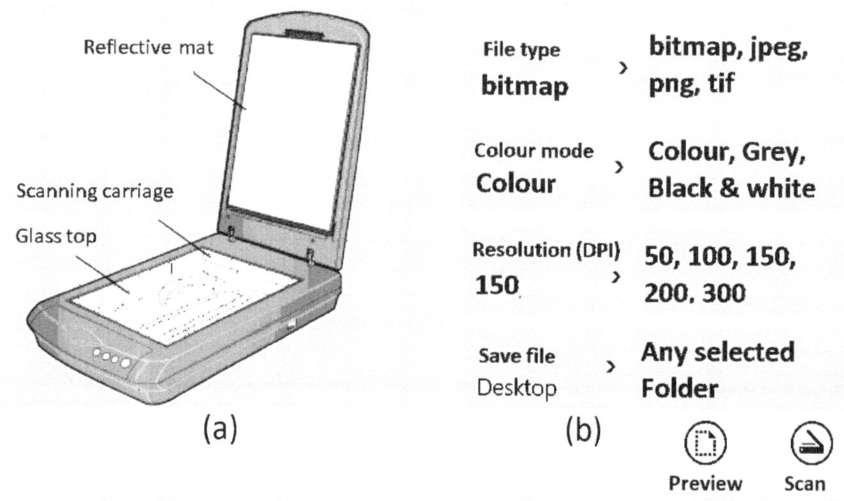

FIGURE 7.1 Scanning the figured motif: (a) parts of the scanner and (b) the dialogue box showing the scanning parameters.

DOI: 10.1201/9781003441205-7

the computer to receive the scanned output. Once the scanner is plugged into the computer, it executes its primary function of scanning the given document or image.

The image on paper or cloth is kept facing the glass top of the scanner. The scanner captures the image of the figure on paper or cloth and converts it into a graphical pixel-picture format. From the scanner memory, the pixelated picture is transferred to the computer memory in a suitable bitmap or pixel file format. This process of creating a bitmap file from the motif using the scanner is called scanning. It is important to note that motifs on paper or cloth kept on the scanner are in vector image form and get converted into raster image form after scanning.

Figure 7.1b shows the four scanning parameters to be decided before starting the scanning. They are (i) resolution, (ii) colour mode, (iii) file type, and (iv) save file. The number of pixels per inch (PPI) or dots per inch (DPI) at which the scanner is set for scanning is called resolution. For the graph designer, the 'pixels per inch' (resolution) is the cloth density or threads per inch (ends per inch and picks per inch). It is possible to set the scanner to any required number of PPI, from 50 to 300. As per the resolution selected, the image gets pixelated and is taken to the memory of the scanner and computer. The image taken for scanning may be in any colour. The scanner has three different colour modes for scanning and saving. They are (i) true colour—24 bitmaps and indexed colour—8 bitmaps, (ii) grey colour, and (iii) black and white (monochrome). As per the colour mode selected, the scanner stores the image in the selected colour mode. It is necessary to select the file type in which the scanned image must be saved on the computer. Bitmap, jpeg, png, and tiff are the different file types used to save the image. After completing the scanning, the image is saved in the required file folder as per the selected file type.

While scanning the motif, the designer can select any one of the colour modes, viz., true colour—24 bitmaps and indexed colour—8 bitmaps, grey colour, and black and white modes in the scanning dialogue box, irrespective of the image colour. Similarly, while saving the image, the designer can select any one of the colour modes in the save dialogue box, irrespective of the colour mode selected in the scanning dialogue box, while scanning. The colour mode selected for saving the image depends on the final image colours required for editing the image. After starting the scanning, a preview of the image is shown on the monitor. If the preview of the image shown is not satisfactory, the position of the paper or the cloth is altered properly, and again, the scanning is started. If the preview of the image shown is satisfactory, it is saved in the required format.

The figured motif taken for scanning may be either in colour as given in Figure 7.2a, or black and white, as given in Figure 7.2b. The size of this motif is 1″ × 1″. Consider that the designer is scanning the figured motif given in Figure 7.2a, drawn on paper with colour outlines. The scanner resolution is kept at 60 PPI. The file type selected is a bitmap, and the colour mode selected is '256 indexed colour'.

The scanner scans the figured motif by dividing it into 60 vertical divisions (60 width density × 1″ width) and 60 horizontal divisions (60 length density × 1″ length). After scanning, the figured motif gets transferred as a raster or pixelated image of 60 × 60 pixels from the scanner memory to the computer memory and opened on the computer. By zooming the pixelated image and opening the graph lines, the figured motif enlarged in the graph size of 60 ends × 60 picks, as shown in Figure 7.2c.

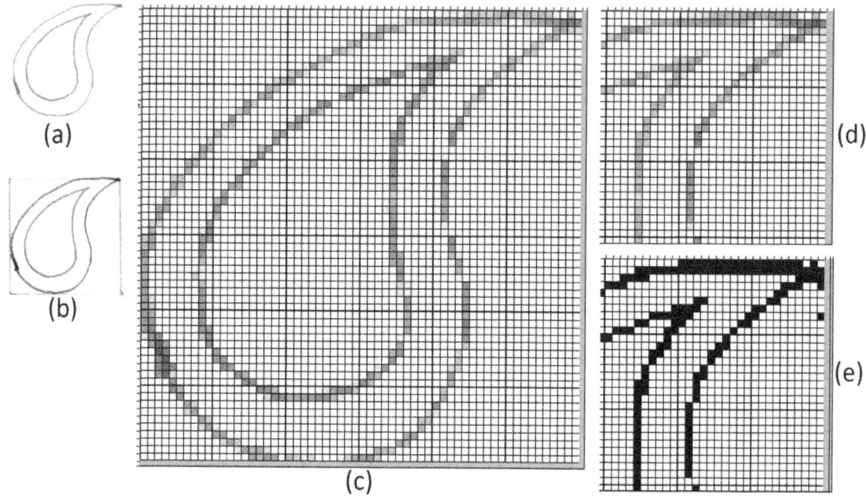

FIGURE 7.2   Scanning and saving by selecting different colour modes: (a) a figured motif in colour, (b) a figured motif in black and white, (c) the colour motif in (a) gets enlarged into the graph size of 60 × 60 in colour after scanning by keeping colour mode while scanning, (d) the black and white motif in (b) gets enlarged into the graph in grey colour after scanning by keeping grey colour mode while scanning, and (e) the black and white motif in (b) gets enlarged into the graph in black and white after scanning by keeping black and white mode while scanning.

The pixels are in different colour tones as per the outline colour of the motif. The file is saved in the folder by selecting any one of the colour modes. According to the colour mode selected while saving, the pixel colour in the graph gets changed. The graph shown in Figure 7.2d is the graph obtained by scanning the black and white motif shown in Figure 7.2b by keeping the grey colour mode while scanning. Figure 7.2e shows the graph obtained by scanning the black and white motif shown in Figure 7.2b by keeping the black and white mode while scanning.

Table 7.1 shows the list of different image colours, colour modes selected while scanning, colour modes selected while saving, and the colours in which the graph file is obtained. The designer must select appropriate options for scanning and saving so that the final pixelated image contains all the details of the image after scanning and can be edited and shape the figure outline properly.

7.2   RESIZING THE SCANNED GRAPH

Some scanners have only a few selective resolution options, like 50, 100, 150, 200, and 300. The designer can't input the resolution as per his requirements other than these options. Hence, while using this type of scanner, the designer can select the next higher resolution for scanning the motif as per the highest cloth density (threads per inch) required and save the graph as it is, as per the size obtained after scanning. Later, when the graph is to be used in a specific size, the designer can reduce the graph size to the required size using the 'resizing option'.

TABLE 7.1

Different Motif Colours, Colour Mode Selected While Scanning, Colour Modes Selected While Saving, and the Colours in Which the Graph File is Obtained

Motif Colour—Reference Figure No.	Colour Mode Selected While Scanning	Colour Mode Displayed on the Monitor	Colour Mode Selected While Saving	The Graph Colour after Saving—Reference Figure No.
Colour (Figure 7.2a)	Colour	Colour	Colour	Colour – (Figure 7.2c)
			Grey colour	Grey colour – (Figure 7.2d)
			Black and White	Black and White – (Figure 7.2e)
Colour (or) Black and White (Figure 7.2a or b)	Grey colour	Grey colour	Colour	Grey colour – (Figure 7.2d)
			Grey colour	
			Black and White	Black and White – (Figure 7.2e)
	Black and White	Black and White	Colour	Black and White – (Figure 7.2e)
			Grey colour	
			Black and White	
Black and White (Figure 7.2b)	Colour	Grey colour	Colour	Grey colour – (Figure 7.2d)
			Grey colour	
			Black and White	Black and White – (Figure 7.2e)

Some advanced scanners have the option to input any required resolution. While using these types of scanners, the designer can scan the motifs in two ways. He can select the highest resolution, scan the motif, and save it. Later, he can resize the graph to the required size. Otherwise, before scanning, he must calculate the resolution considering the motif size and final graph size. He can input the calculated resolution into the scanner, scan the motif, and get the final graph directly as per the required size. In this case, resizing the graph is not necessary.

Furthermore, it is important to understand another aspect of the scanner's resolution. The resolution selected for scanning is common for both length and width directions. For example, when the designer scans the motif of $1'' \times 1''$ size keeping the resolution at 100, the graph size obtained after scanning is 100 ends × 100 picks. This means that the resolution taken for scanning is equal to 'ends per inch (EPI)' and 'picks per inch (PPI)', which are commonly denoted as 'threads per inch'. Hence, eventually, after scanning, all the graphs are obtained with EPI equal to PPI. Hence, it can be used as it is when the cloth is woven with EPI equal to PPI. When the EPI and PPI of the cloth are not equal, the graph obtained after scanning must be resized as per the EPI and PPI.

The common resizing dialogue box of the Computer Aided Figured Graph Designing (CAFGD) software is shown in Figure 7.3a. It has three sets of options, which are three factors: graph size, motif size, and density. The first set of spin boxes is 'total ends' and 'total picks'. It shows the size of the graph in pixels. The next set of spin boxes is 'Width' and 'Height'. It shows the motif size in inches, centimetres, or millimetres. The last set is 'Density-Width' and 'Density-Height' (Density-Length). It shows the resolution or density of the cloth per inch, centimetre, or millimetre. The dialogue box has three lock icons to lock or unlock these three factors. At a time, any one factor is in the lock position, and the other two factors are in the unlock position. By keeping one of the three factors locked, the designer can know the second factor by changing the third factor. The dialogue box also has a chain icon for opening or locking the aspect ratio.

The calculations between the motif size, graph size, and density are given below.

$$\text{Total ends in the graph} = \text{Resolution or Width density (PPI)} \times \text{Motif width in inches}$$

$$\text{Total picks in the graph} = \text{Resolution or Length density (PPI)}$$

$$\times \text{Motif length in inches}$$

From the above equation, the following particulars are calculated:

$$\text{Resolution or Density} - \text{Width (EPI)} = \frac{\text{Total ends in the graph}}{\text{Motif width in inches}}$$

$$\text{Resolution or Density} - \text{Length (PPI)} = \frac{\text{Total picks in the graph}}{\text{Motif length in inches}}$$

$$\text{Motif width in inches} = \frac{\text{Total ends in the graph}}{\text{Resolution or Density} - \text{width (EPI)}}$$

$$\text{Motif length in inches} = \frac{\text{Total picks in the graph}}{\text{Resolution or Density} - \text{length (PPI)}}$$

Consider that the highest thread density of the cloth produced in the region of a designer is 120. The designer has a motif of $1'' \times 1''$ size (Figure 7.2a). He wants to scan the motif and keep it for future use. Therefore, he can scan the image by keeping the scanner resolution at 120 (the highest thread density) and colour mode at 256. After scanning, the motif gets enlarged into the graph size of 120 ends (120 density-width $\times$ $1''$ width) and 120 picks (120 density-length $\times$ $1''$ length) as shown in Figure 7.3a. Keeping this graph open on the screen, when the 'resize' option is selected, the resize dialogue box opens as shown in Figure 7.3b. It opens with the density-width and density-length (resolution) as 100 by default. It is to be noted that it does not open with the 120 resolution taken for scanning. Hence, the dialogue box shows the width as $1.2''$ (120 total ends/100 density-width) and the length as $1.2''$ (120 total picks/100 density-length). All three factors are shown in the dialogue box given in Figure 7.3b. By locking the total ends and total picks, if the designer changes the width and length density from 100 to 120, the width and length will change to 1″ (120 / 120). The 120 ends × 120 picks graph size is saved on the computer by keeping the 256 colour mode in the required folder.

After editing, weave insertion, and punching 120 cards from this graph, the designer can use the card set in a 120 hooks jacquard machine for weaving. If the cloth is woven with 120 EPI × 120 PPI, the figure will be 1″ × 1″ in the cloth, looking like the scanned motif size of 1″ × 1″ (Figure 7.2a). He can also use this card set on a 120 hooks jacquard to weave any other quality cloth having EPI equal to PPI, like 150 × 150, 100 × 100, or 80 × 80. But the figure size in the cloth will not be 1″ × 1″ like the motif. Its size will change according to the cloth density. If the EPI and PPI are 150, the figure size will be 0.8″ × 0.8″ (120 / 150). If the EPI and PPI are 80, the figure size will be 1.5″ × 1.5″ (120 / 80).

Consider that next time the designer requires the graph of the same motif on 60 ends × 60 picks to weave a cloth with 60 EPI and 60 PPI. He can get the 60 ends × 60 picks new graph by resizing (reducing) the 120 ends × 120 picks graph that he has already saved on the computer. The procedure to do this resizing is as follows: (i) open the 120 ends × 120 picks graph already saved (Figure 7.3a); (ii) open the resize dialogue box; (iii) lock the density factor; (iv) open the total ends, total picks, and width-length options; (v) change the total ends from 120 to 60, the total picks from 120 to 60, and lock the graph size factor; and (vi) change the width and length density to 120. Since the density is changed to 120, the programme calculates the width and length and shows them as 0.5″ (60 ends / 120 density-width; 60 picks / 120 density-length). The dialogue box in Figure 7.3c shows all the input as given in the above procedure. Simultaneously, on the computer screen, the graph size is automatically reduced to 60 ends × 60 picks as shown in Figure 7.3d. The number of pixels that were forming the line thickness in the 120 ends × 120 picks graph automatically gets reduced proportionately in the 60 ends × 60 picks graph. By comparing the graph shown in Figure 7.3d with the graph shown in Figure 7.3a, it is possible to understand how the thickness of the line gets reduced proportionately to the reduction of the graph size. The inputs in the dialogue box (Figure 7.3c) reveal that the reduced graph size is 60 ends × 60 picks. Its size will be 0.5″ × 0.5″ in the cloth if it is woven with 120 EPI × 120 PPI.

On another occasion, consider that the designer needs to prepare the graph of the same 1″ × 1″ motif (Figure 7.2a) for weaving the cloth with 120 EPI × 80 PPI. Hence, he needs to prepare the graph as 120 ends (120 EPI × 1″ width) × 80 picks (80 PPI × 1″ length). He can get the new 120 ends × 80 picks graph by resizing (reducing) the 120 ends × 120 picks graph that he has already saved on the computer. The procedure to do this resizing is as follows: (i) open the 120 ends × 120 picks graph already saved (Figure 7.3a); (ii) open the resize dialogue box; (iii) lock the density option, which is 120; (iv) open the total ends-total picks and width-length options; (v) keeping the total ends at 120, change the total picks from 120 to 80; and (vi) lock the graph size factor. Since the density is kept at 120, the programme calculates the width and shows it as 1″ (120 ends / 120 density-width-EPI); the programme also calculates the length and shows it as 0.67″ (80 picks / 120 density-length-PPI). The dialogue box in Figure 7.3e shows all the input as given in the above procedure. Simultaneously, on the computer screen, the graph size is automatically reduced to

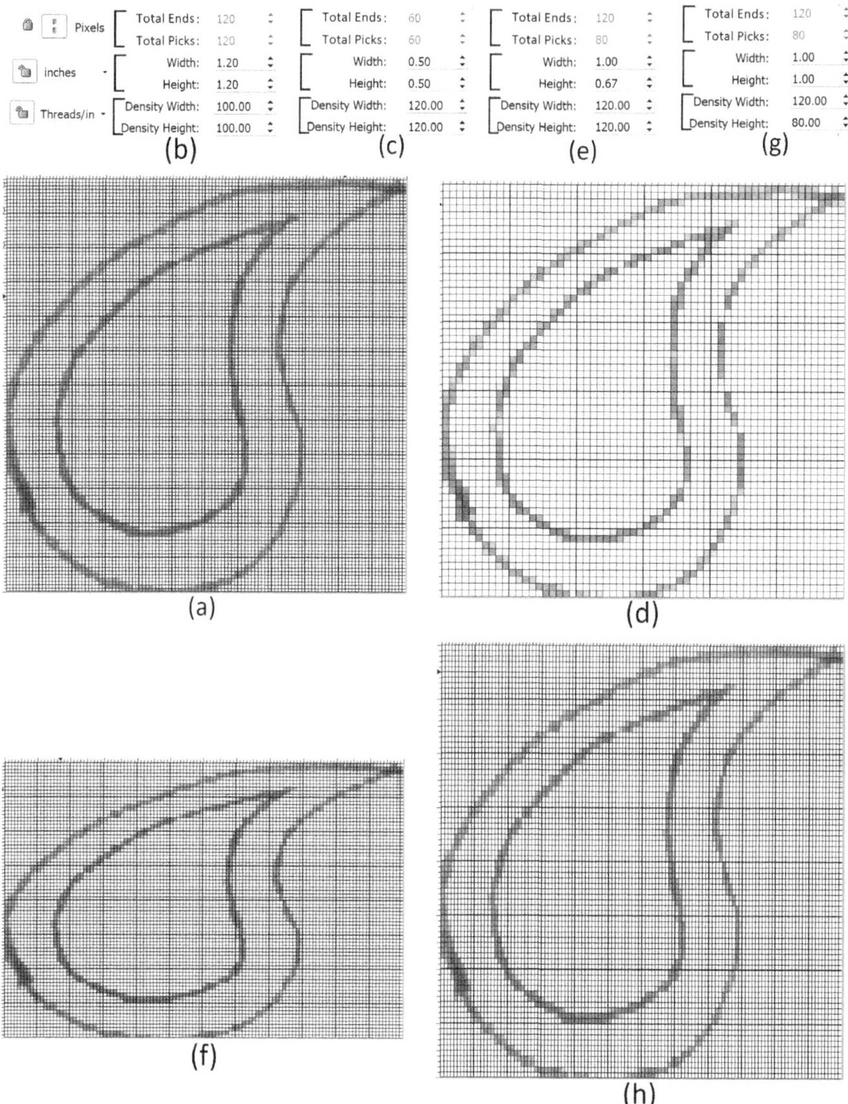

FIGURE 7.3 Different methods of resizing the graph by changing the factors: (a) the 120 × 120 graph obtained by scanning the motif, (b) graph resizing dialogue box showing the resize factors: ends × picks: 120 × 120, width × length: 1.2″ × 1.2″, and density: 100, (c) the dialogue box showing the resize factors changed to ends × picks: 60 × 60, graph width × length: 0.5″ × 0.5″, and density: 120, (d) the 60 × 60 graph obtained by resizing the ends and picks of the graph (a), (e) the dialogue box showing the resize factors changed to ends × picks: 120 × 80, graph width × length: 1.0″ × 0.67″, (f) the 120 × 80 graph, obtained by resizing the ends and picks of the graph (a), (g) the dialogue box showing the resize factors changed to graph width × length: 1.0″ × 1.0″, and density: 120 × 80, and (h) the 120 × 80 graph, obtained by resizing the size of the graph (f).

120 ends × 80 picks as shown in Figure 7.3f. The number of pixels that were forming the line thickness in the 120 ends × 120 picks graph automatically gets reduced proportionately in the 120 ends × 80 picks graph. In this graph, the lengthwise compression of the figure can be observed. This is because the density is kept at 120. To make the graph length proportionate to the graph width, change the length density (PPI) from 120 to 80. The dialogue box in Figure 7.3g shows all the input as given above. By doing so, the length in the dialogue box changes to 1″ (80 picks / 80 density-length-PPI). Simultaneously, on the screen, the 80 picks occupy a length that is equal to the width occupied by the 120 ends as shown in Figure 7.3h. This means that the figure in the graph looks proportionately like the motif shown in Figure 7.2a. By comparing the graph shown in Figure 7.3h with the graph shown in Figure 7.3a, it is possible to understand how the graph size proportionately becomes equal in length and width. Finally, the inputs in the dialogue box (Figure 7.3g) reveal that the reduced graph size is 120 ends × 80 picks. Its size will be 1″ × 1″ in the cloth if it is woven with 120 EPI × 80 PPI.

7.3   REDUCING THE COLOURS IN THE SCANNED GRAPH

The motif given in Figure 7.2a is in a solid colour outline, and Figure 7.2b is in a solid black outline. The graph shown in Figure 7.2c is obtained after scanning Figure 7.2a by keeping the colour mode at 256 colours. The graph shown in Figure 7.2d is obtained after scanning Figure 7.2b (top right) by keeping the colour mode in grey. By scanning, the motif gets enlarged as a graph with the required number of ends and picks. However, the colours of the pixels are as per the colour mode selected while scanning, and the position of pixels is according to the depth and thickness of the lines in the image. The colours and position of the pixels can be observed by comparing the graphs shown in Figure 7.2c and d. It is essential to reduce the number of colours and edit the position of the pixels using various tools and options to bring the outline of the image as per the shape of the individual figure parts required. Thus, reducing the image colours gradually from higher numbers to smaller numbers is called colour reduction. It is easy for the graph designers to do the colour reduction of the figured graph if it is in solid outlines, as seen in Figure 7.2c and d. The outlines in colour pixels or grey pixels can be converted to black and white pixels, as shown in Figure 7.2e, using the colour reduction option (from colour to black and white). In the graph, even though the outline is clearly seen in a single colour, the pixel thickness of the outline is irregular, which must be edited in the next process.

After scanning the figured cloth or motifs with solid colour parts without any outline, it is essential to do the colour reduction gradually from 16.7 million colours to 256 colours to 128 colours to 64 colours to 32 colours, and to any required number of colours between 16 and 2. The designer must carry out the colour reduction until the number of colours in the image gets reduced to as minimum as possible without any loss in the distinct visibility of figure parts from the ground. It will help to edit the outline of the figure parts easily in the next process. Figure 7.4a shows the woven figured cloth of 2″ × 2″ size in three colours, one for the ground and two for the figure. It is scanned by keeping the colour mode at '256 colours' and 200 PPI. The graph

FIGURE 7.4 Reducing the colours in the scanned graph: (a) a woven figured cloth of 2″ × 2″ size in three colours, (b) the scanned image of 400 × 400 graph size, seen on the screen in 2″ × 2″ size with 6 colours, and (c) the resized graph obtained on 100 × 100 from the 400 × 400 graph in six colours.

size obtained on the screen is 400 ends × 400 picks, with the pixels in 256 colours. Then, the colours of the pixels are reduced step by step from 256 to 128 to 64 to 32 to 16 to 6 colours using the colour reduction option. Figure 7.4b shows the stage of the scanned graph on the screen in 2″ × 2″ size with the colours reduced to 6 numbers. Even after reducing the colours to six, the figure parts in Figure 7.4b are as distinctly visible as they are in Figure 7.4a. The 400 ends × 400 picks graph thus obtained after colour reduction is resized to a 100 ends × 100 picks graph using the resize option. The ground colour is also changed to yellow to ensure clear visibility of figure parts for easy editing. Figure 7.4c shows the resized graph to 100 × 100. In the graph, it could be observed that the figure and ground parts are in mixed colour pixels without any clear boundary. This graph must be edited to have the clear boundary of figure parts in a single colour.

7.4   EDITING THE FIGURED GRAPH OUTLINE

After reducing the colours in the scanned graph to the required number and resizing the graph to the desired size, the next process is editing the outline of the figured graph. The scanned image is in many colours according to the colour depth of the figure in the cloth, as seen in the graphs shown in Figure 7.2c, 7.3d, and 7.4c. The designer must edit the figure outline in the graph using computer drawing tools to get the proper shape of all the figure parts as per the design formation. This editing work is the most time-consuming phase in CAFGD. The editing tools will assist the designer in drawing the curves only to certain extent. The graph designer must carefully observe the flow of pixels and change it to achieve the desired curve shape with complete perfection. The general procedure for editing is as follows:

i. Open the graph design to edit. Open the layer option. Open a new transparent, blank graph layer over the graph design layer. The size of the new layer must be equal to the size of the graph design. Figure 7.5a shows the layer option. In the layer option, the graph design opened first for editing is layer 1, and it is indicated as background. The new layer opened is layer 2, and it is indicated as layer 2. The white colour of the new layer 2 is made transparent to see the graph design in the ground layer.

ii. Select a drawing tool like a pencil (T5), Bezier, or curve tool (T16). Select a new colour for the tool other than the colours already present in the graph. The selected colour must also be a contrast colour with the colours present in the graph design. Keep the tool's thickness at one pixel.

iii. Using bezier or curve tool, draw the single-pixel straight line in the new blank layer by connecting two points of the figure curve in the ground layer. By moving the spline points, bend the straight line and make it a curve to exactly overlap with the figure curve.

iv. While drawing, hide and open the graph design layer on the back side now and then to visualize the shape of the outline drawn.

Figure 7.5b shows the curve drawn using the spline tool with a new black colour, in contrast to the blue and red colours present in the graph design (Figure 7.2c). Figure 7.5c shows the curve drawn using the spline tool with a new red colour, in contrast to the colours present in the graph design (Figure 7.4c). After completing the drawing of the outline for all the figure parts, the new layer is saved separately. The graph design

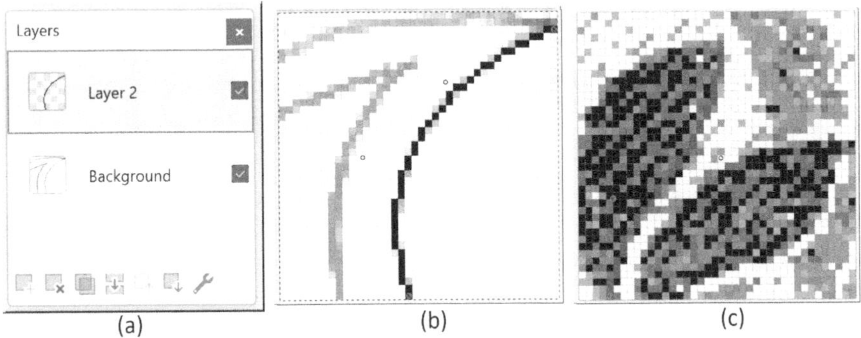

FIGURE 7.5 Editing the figured graph outline using the layer option and the drawing tools: (a) the layer dialogue box with different options, (b) the curve drawn using the spline tool with a new black colour, contrast to the blue and red colours present in the graph design, and (c) the curve drawn using the spline tool with a new red colour, contrast to the colours present in the graph design.

in Figure 7.6a shows the initial black outline drawn using a spline curve over the graph design (60 × 60) shown in Figure 7.3d. It can be observed in the graph that the spline curve tool is not producing the curve line exactly matching the shape of the figure. Hence, it is necessary to edit the outline wherever the shape is irregular. The completely edited graph design on 60 × 60, after observing every pixel is shown in Figure 7.6b. The figure and ground are filled with different colours. Similarly, the graph design in Figure 7.6c shows the initial outline drawn using a spline curve over the graph design (100 × 100) shown in Figure 7.4c. The completely edited graph design after observing every pixel is shown in Figure 7.6d. The figure and ground are filled with different colours. These two edited graph designs (Figure 7.6b and d) are taken for the next process of weave application.

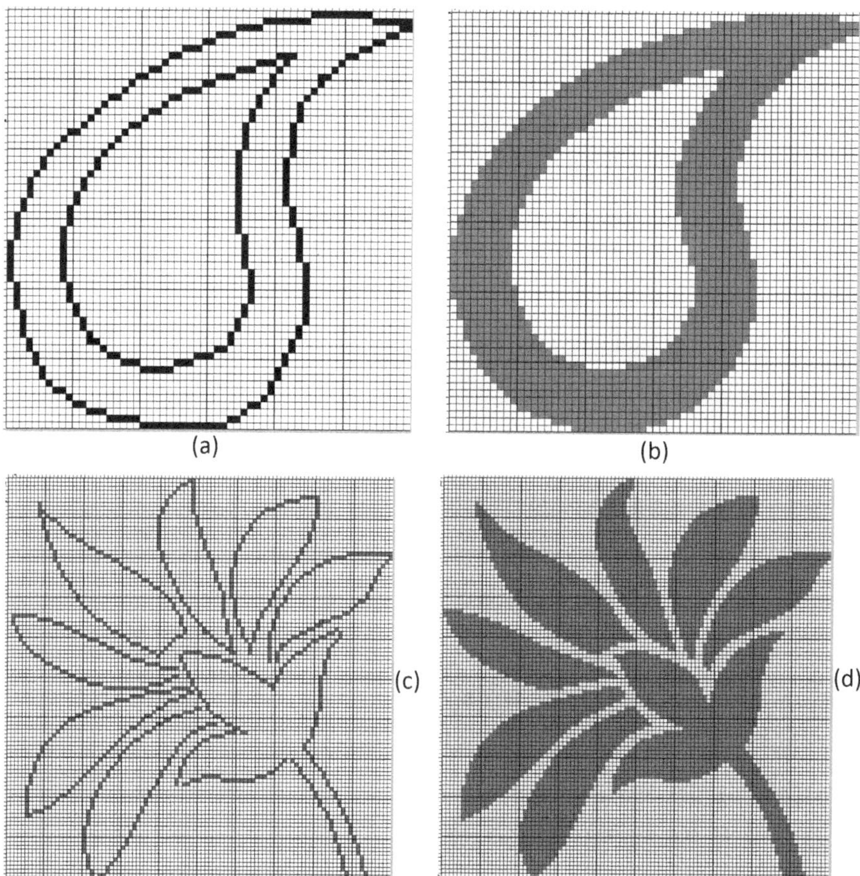

FIGURE 7.6   Graph outlines before editing and after editing: (a) initial black outline drawn using a spline curve over the graph design shown in Figure 7.3d, (b) completely edited graph design after observing every pixel of the graph design in (a) and filled with different colours in figure and ground, (c) initial red outline drawn using a spline curve over the graph design shown in Figure 7.4c, and (d) completely edited graph design after observing every pixel of the graph design in (c) and filled with different colours in figure and ground.

When the motif used for figured graphing is symmetrical and all-over in nature, after scanning the motif, the designer can very well use the rotate and reverse tools along with 'repeat mode' drawing option to visualize the full form of the design and its proper joining when repeating. Figure 7.7 shows the steps for creating different arrangements of designs from the scanned motif using the rotate and reverse tools along with the repeat mode option. Figure 7.7a shows the scanned motif of the simple form created under the diagonal line of one square inch. While scanning, the resolution is set to 300 PPI. Hence, the motif is scanned into 300 ends × 300 picks. Using the colour reduction and graph resizing options, the graph is made into black and white, the size is reduced to 120 × 120, and the ground is filled with yellow, as shown in Figure 7.7b. In the graph, the outlines are scattered pixels, as seen in Figure 7.7c. The outline pixels are not edited at this stage. The figure is reversed diagonally using the rotate and reverse tools. 25% of the figure part is formed as shown in Figure 7.7d. By doing all the reversing, a 100% completed multi-symmetrical figure that looks rectangular is visualized, as shown in Figure 7.7e. Now, this graph size is 240 × 240. Similarly, the second multi-symmetrical figure is developed step by step on 240 × 240 size, as shown in Figure 7.7f–j. This second figure looks diamond-shaped, which is suitable for setting in a half-drop repeat. For a half-drop repeat, there should be sufficient ground surrounding the figure. Hence, the 240 × 240 figure size is reduced to 160 × 160, and then, 40 threads are added on all four sides and made into 240 × 240, as shown in Figure 7.7k. Then, a half-drop repeat is created using the 'cut and shift from the centre' tool. The completed half-drop repeat is shown in Figure 7.7l. The four repeats of the same are shown in Figure 7.7m. In the four repeats shown, it could be observed that the unit figures diagonally touch each other. Consider that the repeat is required without this touching. To achieve it, a slight change in the diagonal reversing figure element is made, as shown in Figure 7.7n. After the multi-reversing is done, the graph size is reduced to 200 × 200 and again made into 240 × 240 by adding 20 threads on all four sides, as shown in Figure 7.7o. The half-drop is done as shown in Figure 7.7p. The four repeats of the figure show that the unit figures are independent without touching each other. Thus, from the simple drawing as shown in Figure 7.7a, f, and h, it is possible to develop four multi-symmetrical figures (Figure 7.7e, j, k, and o) and two half-drop (straight) repeat figures (Figure 7.7l and p). After visualizing all these figures, the selected one, as per the requirement, is taken for editing. Again, while editing, the designer must ensure and judge the formation of shapes and the proper joining of each part very well.

7.5 GRAPH COUNT SETTING

The design paper is divided into vertical and horizontal thick lines to form square blocks. Each block is subdivided into horizontal and vertical spaces. Each horizontal space corresponds to a pick of weft, and each vertical space corresponds to an end of warp and a hook of the jacquard. The number of vertical and horizontal spaces taken in a block is called 'Graph count'. Graph count '10 × 10' means that in a block, there are 10 vertical spaces and 10 horizontal spaces. Graph count '10 × 8' means that in a square block, there are 10 vertical spaces and 8 horizontal spaces. For convenience in the designing and card cutting, the vertical ruling of the paper is arranged to coincide

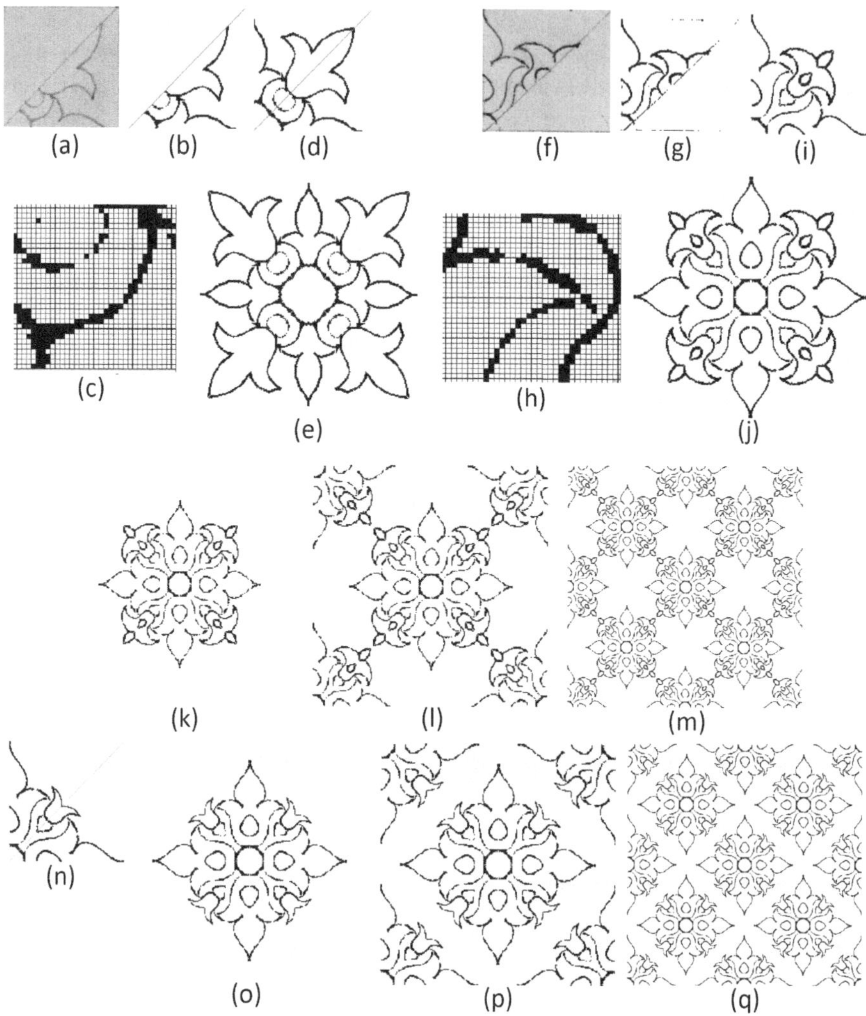

FIGURE 7.7 The steps for creating different types of designs from the scanned motif using the rotate and reverse tools along with the repeat mode option: (a) the scanned motif created under the diagonal line of one square inch, (b) the graph size is reduced to 120 × 120, and the ground is filled with yellow, (c) the graph outlines are in scattered pixels, (d) the diagonally reversed figure is made using the rotate and reverse tools, (e) 100% completed multi-symmetrical figure, (f) the scanned motif created over the diagonal line of one square inch, (g) the graph size is reduced to 120 × 120, and the ground is filled with yellow, (h) the graph outlines are in scattered pixels, (i) the diagonally reversed figure is made using the rotate and reverse tools, (j) 100% completed multi-symmetrical figure, (k) the 240 × 240 figure size is reduced to 160 × 160, and then 40 threads are added on all four sides and made into 240 × 240, (l) half-drop (straight) design repeat is created, (m) four repeats of half-drop design repeat, (n) a slight change in the diagonal reversing figure element is made, (o) 100% completed multi-symmetrical figure of 240 × 240 size is reduced to 200 × 200, 20 threads are added on all four sides and made into 240 × 240, (p) half-drop (straight) design repeat is created, and (q) four repeats of half-drop repeat,

with the arrangement of the jacquard hooks—that is, each large square is divided vertically into as many spaces as there are hooks in a short row of the jacquard. Thus, in the design paper used for an 8-row machine, there are 8 vertical spaces between each pair of thick lines, and for a 12-row jacquard, there are 12 spaces, so that in each case, the number of vertical spaces between the vertical thick lines corresponds to one short row of the card. When the punching is carried long-row-wise of the jacquard, the vertical spaces between the vertical thick lines are taken for every 10 vertical spaces. To retain the correct proportions and shape of figured designs, the number of horizontal spaces in each block must be in the same proportion to the number of vertical spaces as the picks are to the ends per unit space in the finished cloth. Since the number of vertical spaces is fixed by the arrangement of the hooks in the jacquard, it is necessary for the number of horizontal spaces in each square to be varied according to the ratio of picks to ends in the cloth. The editing software has the option to input the required number of vertical spaces (width spaces) and horizontal spaces (length spaces) between the thick lines. Figure 7.8 shows different counts of graph sheets on the monitor and their use in figured graph designing.

For any cloth with equal EPI and PPI, any equal graph count paper is used, viz., 10×10, 8×8, or 12×12, according to the number of hooks per short row of the jacquard machine. Graph count 10×10 (Figure 7.8a) is used for long-row punching; graph count 8×8 (Figure 7.8b) is used for 8-short-row punching; and graph count 12×12 (Figure 7.8c) is used for 12-short-row punching. The proper counts of design paper to suit any given particulars of cloth (EPI, PPI) are derived from the proportion; Ends per Inch: Picks per Inch : : Vertical spaces: Horizontal spaces.

$$\text{That is, Horizontal spaces } = \frac{\text{Vertical spaces}}{\text{Ends per Inch}} \times \text{Picks per Inch}$$

Where vertical spaces are decided as per long-row, 8-row, and 12-row punching.

For example, when the graph is prepared for the cloth having 60 EPI and 48 PPI and long-row punching, the vertical spaces are taken as 10 and the horizontal spaces are calculated as:

$$\text{Horizontal spaces } = \frac{\text{Vertical spaces}}{\text{Ends per Inch}} \times \text{Picks per Inch}$$

$$\text{Horizontal spaces } = \frac{10}{60} \times 48 = 8$$

The graph count required is 10×8, as calculated above and shown in Figure 7.8d.

When the graph is prepared for the cloth having 72 EPI and 54 PPI and 8-row punching, the vertical spaces are taken as 8 and the horizontal spaces are calculated as:

$$\text{Horizontal spaces } = \frac{\text{Vertical spaces}}{\text{Ends per Inch}} \times \text{Picks per Inch}$$

$$\text{Horizontal spaces } = \frac{8}{72} \times 54 = 6$$

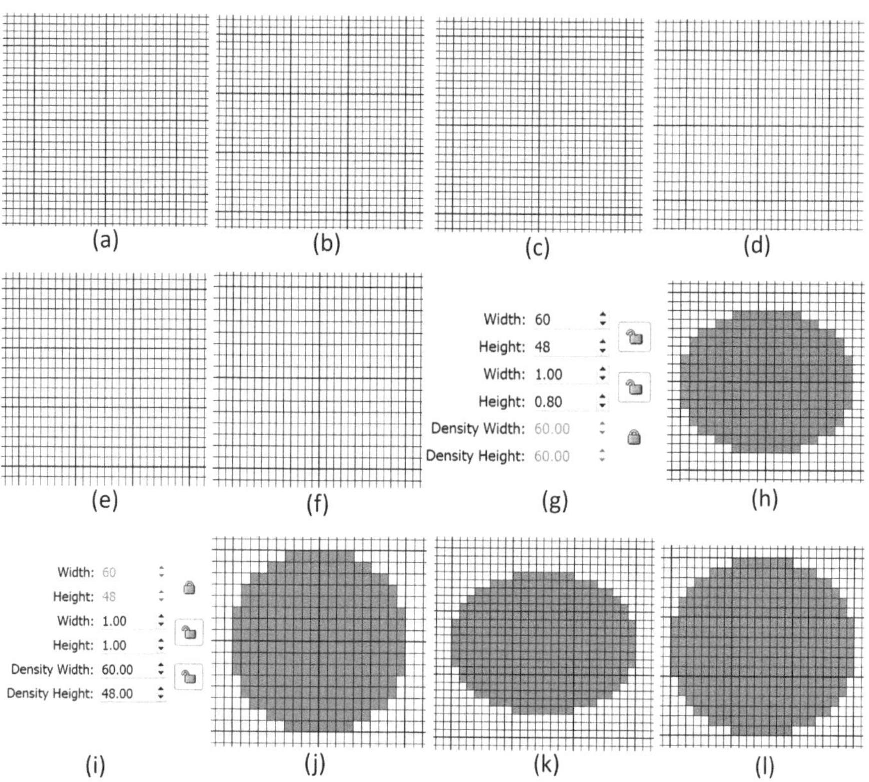

FIGURE 7.8   Graph papers with different graph counts and their use in figured graph designing: (a) graph count—10 × 10, (b) graph count—8 × 8, (c) graph count—12 × 12, (d) graph count—10 × 8, (e) graph count—8 × 6, (f) graph count—12 × 8, (g) the resizing dialogue box set with 60 EPI and 60 PPI, (h) the graph count is set in 10 × 10 and the ellipse is drawn on 20 ends × 16 picks, (i) the resizing dialogue box set with 60 EPI and 48 PPI, (j) the graph count is set in 10 × 8 and the ellipse is drawn on 20 ends × 16 picks looks like a circle (k) the graph count set in 8 × 8 and the ellipse is drawn on 24 ends × 18 picks, and (l) the graph count is set in 8 × 6 and the ellipse drawn in 24 ends × 18 picks looks like a circle.

The graph count required is 8 × 6, as calculated above and shown in Figure 7.8e.

When the graph is prepared for the cloth having 96 EPI and 64 PPI and 12-row punching, the vertical spaces are taken as 12 and the horizontal spaces are calculated as:

$$\text{Horizontal spaces} = \frac{\text{Vertical spaces}}{\text{Ends per Inch}} \times \text{Picks per Inch}$$

$$\text{Horizontal spaces} = \frac{12}{96} \times 64 = 8$$

The graph count required is 12 × 8, as calculated above and shown in Figure 7.8f.

For example, when the EPI is equal to the PPI, say 60, for a small circle of 0.33″ × 0.33″ in size, the graph is prepared on 20 ends × 20 picks. But when the EPI is 60 and the PPI is 48, for a circle of the same size, the graph should be on 20 ends (0.33″ × 60) × 16 picks (0.33″ × 48). If this circle is drawn by keeping EPI at 60 and PPI at 60 in the resize dialogue box as shown in Figure 7.8g and taking the graph count as 10 × 10, it looks like an ellipse, as shown in Figure 7.8h. But if the PPI is changed to 48, as shown in Figure 7.8i, and the graph count is 10 × 8, the circle drawn will look like a circle, as shown in Figure 7.8j.

For another example, when the EPI is equal to the PPI, say 48, for a small circle of 0.5″ × 0.5″ size, the graph is prepared on 24 ends × 24 picks. But when the EPI is 48 and the PPI is 36, for a circle of the same size, the graph should be 24 ends (0.5″ × 48) × 18 picks (0.5″ × 36). If this circle is drawn by keeping EPI at 48 and PPI at 48 in the resize dialogue box and taking the graph count as 8 × 8, it looks like an ellipse, as shown in Figure 7.8k. But if the PPI is changed to 36 and the graph count is 8 × 6, the circle drawn will look like a circle, as shown in Figure 7.8l.

7.6 MOTIF SIZE ON THE PAPER AND THE LOOM

It is pertinent that once the harness is built in the jacquard, the repeat width of the harness or ends is fixed. This result in the repeat width of the figure produced on the cloth (on the loom) is also fixed and cannot be changed. Hence, when a new motif is drawn for the jacquard loom, which is already built with harnesses, the repeat width on the loom is calculated first. The new motif is drawn keeping in mind that the width taken on the paper is equal to the repeat width on the loom. The length of the motif taken can be any size according to the requirement or designer's choice. Alternatively, the motif already available in a certain width and length is taken for graphing to weave on the available loom already built with a definite harness width per repeat. In this case, sometimes the motif width on the paper may be equal to the repeat width on the loom. But many times, the motif width may not be equal to the repeat width on the loom. It may be more, or it may be less. Thus, there are three conditions that occur. The first one is that the motif size on the paper is equal to the repeat width on the loom. The second one is that the motif size on the paper is larger than the design repeat size on the loom. The third one is that the motif size on the paper is smaller than the repeat width on the loom. But when these motifs of different widths are woven on the loom, the figure width produced on the cloth will be equal to the repeat width of the harness on the loom, invariably to the motif width on the paper. It is important that, in all three cases, the width and length of the motif on paper and the width and length of the design produced on the cloth be proportionate so that the design elements on the cloth are proportionate to the design elements on the motif without any distortion. The length of motif to be drawn on paper is purely at the discretion of the designer which depends on the character of the motif and will have the following three options: (i) motif length may be same as the motif width; (ii) motif length may be more than the motif width; and (iii) motif length may be less than the motif width. Hence, the designer has to know how to calculate the ends and picks of the graph to be taken, considering the motif size on the paper as well as the harness repeat size on the loom. The different calculations involved are given below with examples.

Consider that a motif of 6″ width and 4″ length on paper, as shown in Figure 7.9a, is selected for weaving. It is used for three jacquard looms set to weave fabrics of three different thread densities (EPI and PPI). The calculations involved in transferring the motif into graph for these three looms are as follows:

i. The motif of 6″ width and 4″ length on paper (Figure 7.9a) is selected for weaving on the loom set with Jacquard capacity = 480 hooks; EPI = 80; PPI = 80.

    Harness tie −1 Harness = 1 End; So, ends per repeat = 480.

    Hence, the width of the design repeat on the loom = (Ends per repeat / EPI) = (480 / 80) = 6″.

    Width of the motif on paper = Width of the design repeat on loom = 6″.

$$\text{Length of design on loom} = \text{Length of motif on paper} \times \frac{\text{Width of repeat on loom}}{\text{Width of motif on paper}}$$

$$\text{Length of design on loom} = 4'' \times \frac{6''}{6''} = 4''$$

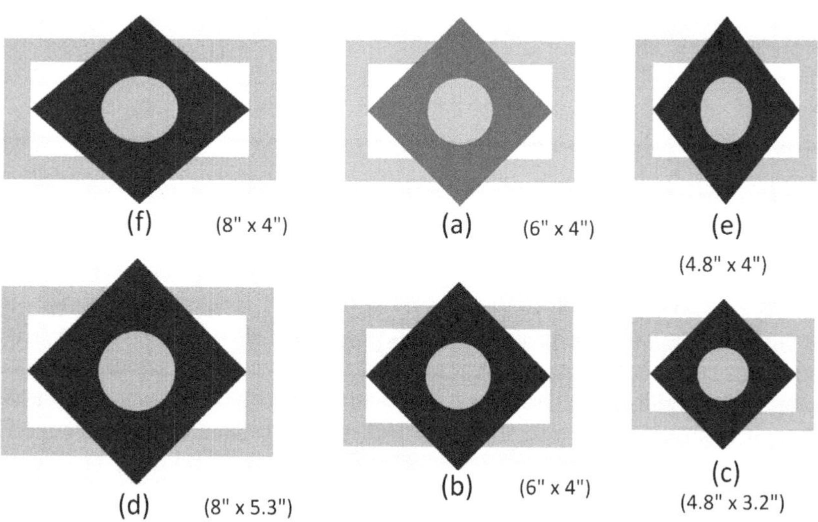

FIGURE 7.9 Motif size on paper and repeat size on the loom: (a) a motif on paper in 6″ × 4″ size, (b) design repeat size on the cloth/loom (6″ × 4″) equal to motif size, (c) design repeat size on the cloth/loom (4.8″ × 3.2″) proportionately smaller than motif size, (d) design repeat size on the cloth/loom (8″ × 5.3″) proportionately bigger than motif size, (e) design repeat size on the cloth/loom (4.8″ × 4″) and with design elements compressed in width way, and (f) design repeat size on the cloth/loom (8″ × 4″) with design elements elongated in width way.

Hence, the *length of the motif on paper= the length of the repeat on loom = 4″.*

The size of the motif on paper = 6″ × 4″ and the repeat width on loom = 6″ × 4″, as shown in Figure 7.9b.

Total ends in the graph = Ends per repeat = 480.

Total picks in the graph = (Length of the design in inches × PPI) = 4″ × 80 = 320

Total Graph size = 480 Ends × 320 Picks. Total cards punched = 320.

Graph count = EPI : PPI :: Vertical spaces : Horizontal spaces = 80 : 80 = 10 : 10 or 8 : 8.

By comparing Figure 7.9a and b, it is noted that the design elements are equal in size both in motif and on the loom.

ii. The same motif of 6″ width and 4″ length on paper (Figure 7.9a) is selected for weaving on another loom set with Jacquard capacity = 480; EPI = 100; PPI = 80.

Harness tie = 1 Harness-1 End; So, ends per repeat = 480.

Hence, the repeat width on the loom = (Ends per repeat / EPI) = (480 / 100) = 4.8″.

The width of the motif on paper = 6″; the repeat width on the loom = 4.8″

That is, the width of the motif on paper is not equal to the repeat width on the loom.

The 6″ width of motif on paper will be 4.8″ repeat width on cloth. (reduction in width because of increase in EPI)

Hence, the length of the motif should also be reduced proportionately.

$$\text{Length of design on loom} = \text{Length of motif on paper} \times \frac{\text{Width of repeat on loom}}{\text{Width of motif on paper}}$$

$$\text{Length of design on loom} = 4\,'' \times \frac{4.8''}{6''} = 3.2''$$

The 4″ length of motif on paper must be 3.2″ length of repeat on the loom (the reduction in length is proportionate to the reduction in width)

The size of motif on paper = 6″ × 4″ and the design repeat on loom = 4.8″ × 3.2″. The motif size is proportionately reduced in the cloth, as shown in Figure 7.9c.

Total ends in the graph = Ends per repeat = 480.

Total picks in the graph = (Length of the design in inches × PPI) = 3.2″ × 80 = 256

Total Graph size = 480 Ends × 256 Picks. Total cards punched = 256.

Graph count = EPI: PPI :: Vertical spaces: Horizontal spaces = 100 : 80 = 10 : 8, 8 : 6 or 8 : 7.

By comparing Figure 7.9a and c, it is noted that the design elements of the motif on the paper are proportionately reduced to the design on the loom. To get the length proportionate to the width on loom, it is necessary to have 256 cards/picks instead of 320 cards/picks as in the case of first loom.

iii. The motif of 6″ width and 4″ length on the paper is also used in the third loom set with Jacquard capacity = 480; EPI = 60; PPI = 80.

Harness tie = 1 Harness-1 End; so, ends per repeat = 480.

Hence, the width of the design repeat on the loom = (Ends per repeat / EPI) = (480 / 60) = 8″.

The width of the motif on paper = 6″; the width of the design repeat on the loom = 8″

The width of the motif on paper is not equal to the width of the design repeat on the loom.

The 6″ width of the motif on the paper will be 8″ repeat width on the loom. (increase in width because of reduction in EPI)

Hence, the length of motif should also be increased proportionately.

$$\text{Length of design on loom} = \text{Length of motif on paper} \times \frac{\text{Width of repeat on loom}}{\text{Width of motif on paper}}$$

$$\text{Length of design on loom} = 4'' \times \frac{8''}{6''} = 5.3''$$

The 4″ length of the motif on paper must be 5.3″ length of repeat on the loom (the increase in length is proportionate to the increase in width).

The size of motif on paper = 6″ × 4″ and the design repeat on loom = 8″ × 5.3″. The motif size is proportionately increased in the cloth, as shown in Figure 7.9d.

Total ends in the graph = Ends per repeat = 480.

Total picks in the graph = (Length of the design in inches × PPI) = 5.3″ × 80 = 424

Total Graph size = 480 Ends × 424 Picks. Total cards punched = 424.

Graph count = EPI : PPI :: Vertical spaces : Horizontal spaces = 60 : 80 = 10 : 13 or 8 : 11.

By comparing Figure 7.8a and d, it is noted that the sizes of design elements are proportionately increased from the motif on the paper to the design on the loom. To get the length proportionate to the width on loom, it is necessary to have 424 cards/picks instead of 320 cards/picks as in the case of first loom.

Assume that the 320 cards punched for the first loom (80 ends × 80 picks), by mistake, are used in the second loom (100 ends × 80 picks); then, the design repeat on the loom will be 4.8″ × 4.0″ as shown in Figure 7.9e. The size of motif on paper = 6″ × 4″ and the design repeat on loom = 4.8″ × 4.0″. That is, only the motif width is reduced but the length remains the same as shown in Figure 7.9e. By comparing Figure 7.9a and e, it is noted that the design elements are compressed in the width way which is not desired (the circle becomes oblong).

Assume that the 320 cards punched for the first loom (80 ends × 80 picks), by mistake, are used in the third loom (60 × 80). Then the design repeat on the loom will be 8.0″ × 4.0″ as shown in Figure 7.9f. The size of motif on paper = 6″ × 4″ and the design repeat on loom = 8.0″ × 4.0″. That is, only the motif width is increased but the length remains the same as shown in Figure 7.9f. By comparing Figure 7.9a and f, it is noted that the design elements are elongated in the width way which is not desired (the circle becomes ellipse).

7.7 WEAVE MAPPING

In the graph designing of figured fabrics, after editing the scanned motif outline perfectly, the weave marks are applied as per the variety of fabric to be produced. In manual graph designing, the designer selects the type of weave marks to be given and applies them to the selected parts. While applying the binding weave marks manually, the designer restricts or controls where to give and not give binding marks as per the graph variety produced and float length allowed. In computer-aided figured graph designing, it is possible to assign different weaves for different colours (for one colour, one weave). Hence, it is essential to fill up different parts of the figure with different colours as per the weaves to be applied. After filling up with different colours, keeping the above aspect in mind, it is possible to apply different weaves for different colours. This process of applying different weaves to different colour parts of the design as per the cloth structure is called 'weave mapping'.

The general steps to be followed for applying the weave marks are (i) apply different colours to the different graph parts as per the different weaves to be applied, (ii) create one repeat of the weaves to be applied and save them, (iii) take the brush option and assign the weave to apply, (iv) keep the graph count equal to the repeat size of the weave, and (v) take the colour fill tool and click on the bottom left pixel (x1, y1) of any one square or rectangle thick line block, in the colour area of the graph on which the weave has to be applied. By doing so, the weave assigned to the brush gets applied completely over the colour on which the fill tool is clicked. The graph count used is equal to the repeat size of the weave, and clicking the fill tool on the bottom left leads to the application of the weave to repeat similarly in different parts of the graph with the same colour.

For example, a part of the edited figured graph on 40 × 40 is shown in Figure 7.10a. It is applied with three different colours to map/apply three different weaves. The yellow colour is for the ground to apply the 8 × 8 sateen weave. The red colour is for the figure to apply the 5 × 5 satin weave. The light blue colour is for the ground area inside the red figure area to apply 4 × 4 twill weave (1 up, 3 down). The initial count of the figured graph is 10 × 10.

The process of applying these three weaves to three different colours is carried out one by one. The process of applying the 8 × 8 sateen weave to the ground in yellow is carried out with the following steps: change the graph count from 10 × 10 to 8 × 8, create one repeat of the 8 × 8 sateen weave (using blue colour) on any one of the 8 × 8 thick square blocks on the yellow colour area of the graph, and select the weave repeat as shown in Figure 7.10b. Transfer this selected weave layer to the brush by clicking the layer to brush option. Select the weave appearing in the brush option, check the brush option, and check the whole image option as shown in Figure 7.10c. Take the colour fill tool and click on the bottom left pixel of any 8 × 8 thick square

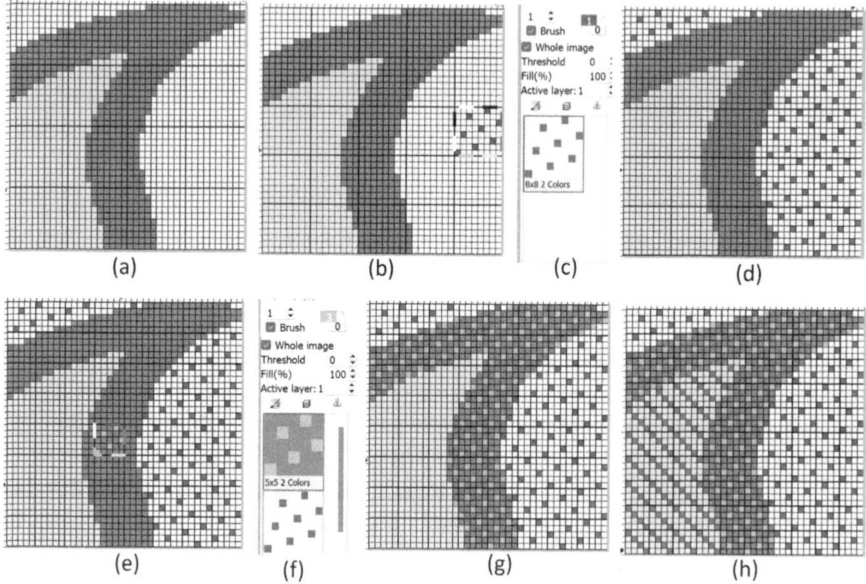

FIGURE 7.10   Weave mapping: (a) edited figured graph of 40 × 40 with three different colours to apply three different weaves, (b) 8 × 8 sateen weave is created on yellow and copied as a layer to transfer as a brush, (c) brush option showing the 8 × 8 sateen weave in selection with brush and whole image options checked, (d) edited graph applied with 8 × 8 sateen weave, (e) 5 × 5 sateen weave is created on red and copied as a layer to transfer as a brush, (f) brush option showing the 5 × 5 sateen weave in selection with brush and whole image options checked, (g) edited graph applied with 5 × 5 sateen weave, and (h) edited graph applied with 4 × 4 twill weave.

line block on the yellow colour area of the graph on which this weave has to be applied. By doing so, the 8 × 8 sateen weave assigned to the brush gets applied on all the yellow colour areas in the graph, as shown in Figure 7.10d.

The work of applying the 5 × 5 sateen weave to the figure in red is carried out with the following steps: change the graph count from 8 × 8 to 5 × 5, create one repeat of the 5 × 5 sateen weave (using green colour) on any one of the 5 × 5 thick square line blocks in the red colour area of the graph, and select the weave repeat as shown in Figure 7.10e. Transfer this selected weave layer to the brush by clicking the layer to brush option. Select the weave appearing below the brush option, check the brush option, and check the whole image option as shown in Figure 7.10f. Take the colour fill tool and click on the bottom left pixel of any 5 × 5 thick square line block in the red colour area of the graph on which this weave has to be applied. By doing so, the 5 × 5 sateen weave assigned to the brush gets applied on all the red colour areas in the graph, as shown in Figure 7.10g. Similarly, the third weave (1 up, 3 down twill) is also mapped on the light blue colour area by changing the graph count from 5 × 5 to 4 × 4, as shown in Figure 7.10h.

Though the warp and weft face weaves are applied in the figured graph as shown in Figure 7.10h, it is important to note that the weave marks get applied on all the colour pixels, including in the corners of the stepping and also over the shorter floats of ends and picks. Hence, it is necessary to apply the weave marks only in the places where they are required and not in the places where they are not required to apply them by following the different methodology called 'controlled application of weave marks'. The different weaves applied in the weave mapping process are in different colours. Hence, after mapping, all the colours that indicate warp up in all the weaves are converted into warp colour used in the fabric. Similarly, all the colours that indicate warp down (weft up) in all the weaves are converted into weft colour used in the fabric.

BIBLIOGRAPHY

1. ArahPaint. (2018). *User's Manual*. Arahne Software Company in Ljubljana, Slovenia.
2. Auto Tex 2000. (2005). *User's Manual*. PLC Consulting Company, New Delhi.
3. CadVantage Win Jacquard. (2001). *User's Manual*. Teckmen Systems Pvt. Ltd., Pune.
4. Grosicki, Z. J. (2004). Counts of design paper. In *Watson's Textile Design and Colour* (pp. 197–199). Woodhead Publishing Limited.
5. Textronics. (2000).*User's Manual of Design*. Jacquard Textronics Design Systems (I) Pvt. Ltd., Mumbai.

8 Weave Creation, Float, and Float Control

8.1 WEAVE CREATION USING CAFGD SOFTWARE

After completing the preparation of the figured graph with the required outline as per the motif, different weaves are applied on different figure parts as per the fabric variety for which the graph is prepared. Applying the appropriate weave must enhance the effect of the fabric without affecting the design character. Hence, the designers need to have as many weaves as possible on the computer that they can readily use wherever and whenever required. The CAFGD software is useful for the figured graph designers to create the required weave and save it as a file in a file folder. All the file folders thus prepared together are called 'Weave Library'. From this weave library folder, the required weave file is selected and as a brush option for applying it to the different parts of the jacquard graph. The software has different tools and options to create simple and compound weaves easily. According to the simple or compound weave to be created, the designer can select the appropriate tools and options.

8.2 SIMPLE WEAVES CREATION

8.2.1 TWILL AND SATEEN WEAVES

The software has a 'Titling option' to create twill and sateen weaves. Figure 8.1 shows how to get 8 × 8 twill and sateen weaves by drawing a simple vertical line and using the tilting option. First, open a graph of 8 × 8 size. Next, draw a vertical line on the first end, as shown in Figure 8.1a. It is modified to different weaves with consecutive clicks on the tilting option. Now, click the 'tilting option' one time. The eight pixels of the vertical line move in 1 out-1 up order, forming a 1 up 7 down twill line running left to right as shown in Figure 8.1b. After that, for the second click of the 'tilting option', the pixels move in a 2 out-1 up order, as shown in Figure 8.1c. Likewise, the move increases to 3 out-1 up for the third click, to 4 out-1 up for the fourth click, and so on, as seen in Figure 8.1d–g. For the seventh click, the pixels move as seen in Figure 8.1h, resulting in a 1 up 7 down twill line running right to left diagonally. For the eighth click, the result is a vertical line on the first end as shown in Figure 8.1i, which is similar to the line in Figure 8.1a. Out of these seven weaves, the weaves in Figure 8.1b and h are twill weaves. Out of the remaining five weaves seen from Figure 8.1c–g, the weave in Figure 8.1d is the correct regular sateen weave with 3 out-1 up move, and the weave in Figure 8.1f is the correct regular sateen weave with 5 out-1 up move. The weaves in Figure 8.1c, e and g are not the sateen weaves. These weaves can be used as binding weaves for the extra weft. All seven weaves are saved with corresponding file names while doing each click.

DOI: 10.1201/9781003441205-8

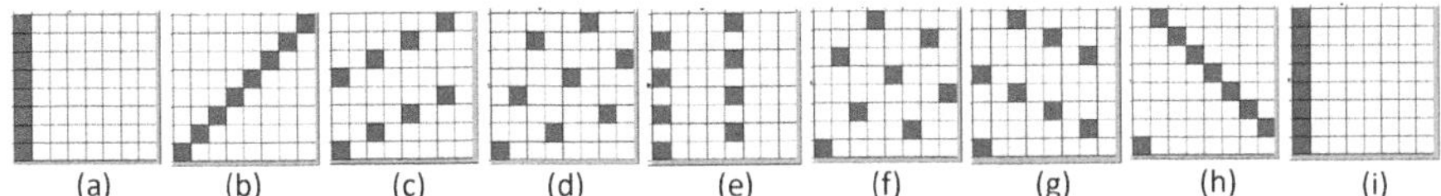

FIGURE 8.1 Twill and sateen weaves creation using the 'tilting option' of CAFGD: (a) a vertical line drawn on the first end, (b) 1 up-7 down twill weave (left to right) obtained by the first click, (c) extra weft binding weave (2 out-1 up) obtained by the second click, (d) sateen weave (3 out-1 up) obtained by the third click, (e) extra weft binding weave (4 out-1 up) obtained by the fourth click, (f) sateen weave (5 out-1 up) obtained by the fifth click, (g) extra weft binding weave (6 out-1 up) obtained by the sixth click, (h) 1 up-7 down twill weave (right to left) obtained by the seventh click, and (i) getting again the vertical line on the first end by the eighth click.

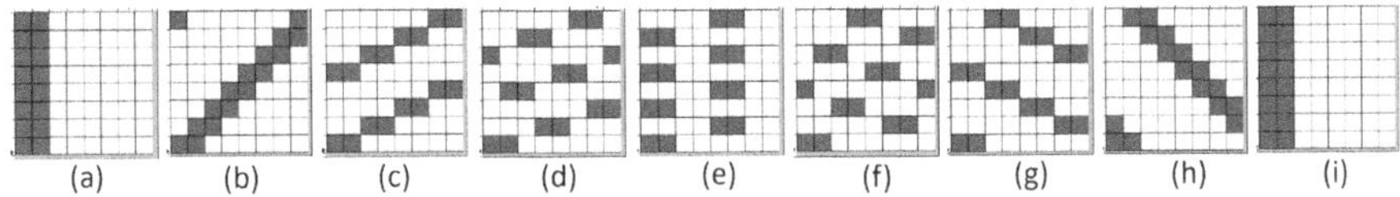

FIGURE 8.2 Method of creating 2 up 6 down twill and sateen weaves of 2 marks. (a) vertical lines drawn on the first two ends, (b) 2 up-6 down twill weave (left to right) obtained by the first click, (c) extra weft binding weave (2 out-1 up) obtained by the second click, (d) sateen weave (3 out-1 up) obtained by the third click, (e) extra weft binding weave (4 out-1 up) obtained by the fourth click, (f) sateen weave (5 out-1 up) obtained by the fifth click, (g) extra weft binding weave (6 out-1 up) obtained by the sixth click, (h) 2 up-6 down twill weave (right to left) obtained by the seventh click, and (i) getting again the vertical line on the first two ends by the eighth click.

Another example is shown in Figure 8.2. Two vertical lines are drawn on the first two ends of 8 × 8, as shown in Figure 8.2a. The 2 up-6 down twill weave and sateen weaves of 2 marks obtained by each click are shown in Figure 8.2b–h. For the eighth click, the result is two vertical lines on the first two ends as shown in Figure 8.2i, which is similar to the two vertical lines in Figure 8.2a. Out of these seven weaves, the weaves in Figure 8.2b and h are 2 up-6 down twill weaves with left-to-right and right-to-left twill lines, respectively. Out of the remaining five weaves seen from Figure 8.2c–g, the weave in Figure 8.2d is the correct regular sateen weave of 2 marks with 3 out-1 up move, and the weave in Figure 8.2f is the correct regular sateen weave of 2 marks with 5 out-1 up move. The weaves in Figure 8.2c, e and g are not the sateen weaves. But the weaves in Figure 8.2c and g are 2-mark stepped lines on 8 × 4 in left-to-right and right-to-left directions, respectively. These weaves can be used as binding weaves for the extra weft with two ends stitching. All seven weaves are saved with corresponding file names while doing each click.

8.2.2 Wavy Twill and Diamond Weaves

Figure 8.3 shows how to create wavy twills and diamond weaves using 'copy', 'reverse', and 'paste' options. Figure 8.3a shows the vertical lines drawn on the first, second, fourth, tenth, and twelfth ends of 12 × 12 to get a twill weave of 3 up-1

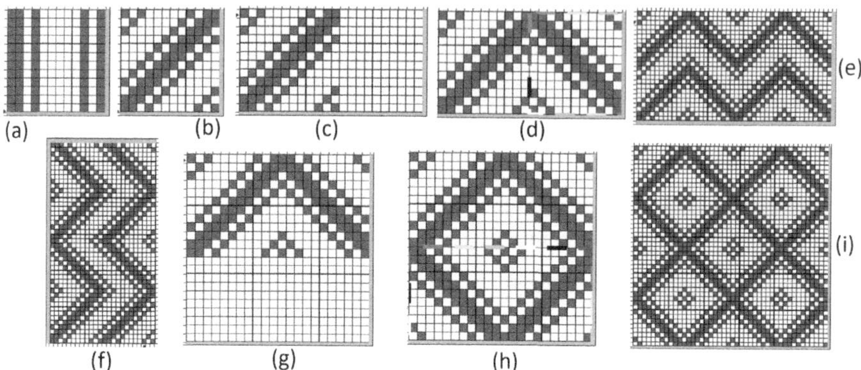

FIGURE 8.3 Wavy twill and diamond weaves creation from twill using copy, reverse, and paste options of CAFGD: (a) vertical lines drawn on 12 × 12, (b) twill weave obtained by the first click of the tilting option, (c) opening 22 × 12 graph; copying, and pasting the twill weave on the left, (d) wavy twill across weave obtained on 22 × 12 after copying, reversing, and pasting the twill, (e) four repeats of the wavy twill across weave, (f) wavy twill along weave obtained by rotating the wavy twill across weave by 90°, (g) opening 22 × 22 graph; copying, and pasting the wavy twill across on the top, (h) diamond weave obtained on 22 × 22 after copying, reversing, and pasting, and (i) four repeats of the diamond weave.

down-1 up-5 down-1 up-1 down. Figure 8.3b shows the twill weave obtained by the first click of the tilting option. To create a 'wavy twill across' weave, open the graph of 22 ends [(12 ends × 2) − (2)] and 12 picks. Copy and paste the twill weave of 12 × 12 on the left side of the 22 × 12 graph, as shown in Figure 8.3c. The 10 middle ends of the twill weave are selected, copied, and reversed in the end-way. The reversed twill weave is moved and pasted over the ten blank ends on the right side. This results in the formation of a 'wavy twill across' weave with the twill reversing on the first and twelfth ends, as shown in Figure 8.3d. The correctness of the weave repeat is verified by taking 2 end-way repeats and 2 pick-way repeats in the 'repeat mode' option, as shown in Figure 8.3e.

The 'wavy twill along' weave, as seen in Figure 8.3f, is obtained by rotating the 'wavy twill across weave' to 90°. The diamond weave is constructed from the 'wavy twill across' weave on 22 × 12. For this, open the graph of 22 ends [(12 ends × 2) − (2)] and 22 picks [(12 picks × 2) − (2)]. Copy and paste the 'wavy twill across' weave on the top side of the 22 × 22 graph as shown in Figure 8.3g. Next, the ten middle picks of the 'wavy twill across' weave are selected, copied, and reversed in the pick-way. Then, the reversed 'wavy twill across' weave is moved and pasted over ten blank picks at the bottom, resulting in the formation of the diamond weave with the wavy twill across weave reversing on the first and twelfth pick, as shown in Figure 8.3h. The correctness of the weave repeat is verified by taking 2 end-way repeats and 2 pick-way repeats in the 'repeat mode' option, as shown in Figure 8.3i.

8.3 COMPOUND WEAVES CREATION

8.3.1 Warp and Weft-Backed Cloth Weaves

The software has a set of two options, namely 'Divide vertical section' and 'Merge vertical section'. Similarly, it has another set of two options, namely 'Divide horizontal section' and 'Merge horizontal section'. These options are useful for creating compound weaves like backed cloth and double cloth. Figure 8.4a and b shows how the 'divide vertical section' and 'merge vertical section' options work to derive the compound weave. Figure 8.4a shows a diamond graph on 32 × 16. The odd ends are in red, and the even ends are in blue. By keeping the sections as two (x = 2) and clicking the 'divide vertical section' option, the graph gets converted as shown in Figure 8.4b. In the graph, the odd red ends are grouped and made into the first section of 16 × 16, and the even blue ends are grouped and made into the second section of 16 × 16. Again, by clicking the 'merge vertical option', the graph gets converted back, as shown in Figure 8.4a. These two options are used to create a warp-backed weave.

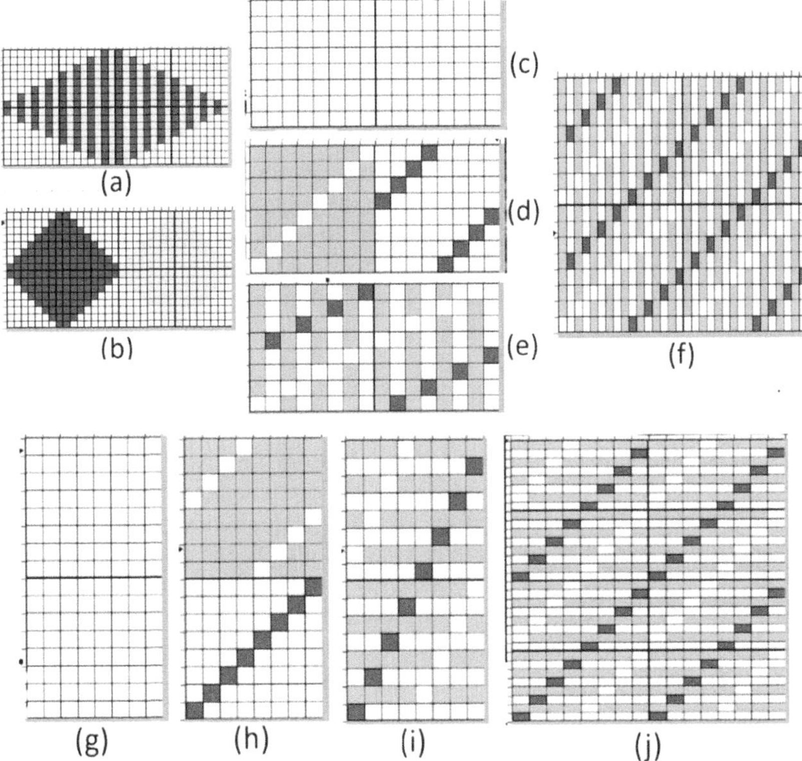

FIGURE 8.4 Warp-backed and weft-backed cloth weaves creation, using the divide and merge options of CAFGD. Figure 8.4a and b shows the function of divide and merge options. Figure 8.4c–f shows how to create a warp-backed weave using the 'divide vertical' and 'merge vertical' options. Figure 8.4g–j shows how to create a weft-backed weave using the 'divide horizontal' and 'merge horizontal' options.

For example, consider that it is required to create a warp-backed weave on 16×8 by combining a warp face twill on 8×8 (7 up-1 down) and a weft face twill on 8×8 (1 up-7 down) in a 1:1 end-way. To achieve this, first, take a graph of 16×8, as shown in Figure 8.4c. Click on the 'divide vertical section' option by taking the vertical section (x) as 2. By doing this, the odd ends in the graph are grouped as the first section of eight ends on the left side, and the even ends in the graph are grouped as the second section of eight ends on the right side, as described by the graphs in Figure 8.4a and b. Next, mark a 7 up-1 down twill in the left section of eight ends, as shown by the green colour (up) in Figure 8.4d. Mark a 1 up-7 down twill in the right section of eight ends, as shown by the red colour (up) in Figure 8.4d. Then, click the 'merge vertical section' option. By clicking, the ends in two sections get merged in a 1:1 order, resulting in a warp-backed weave as shown in Figure 8.4e.

After that, check the repeating of the weave by taking four repeats. In the warp-backed weave, mostly, the ends per inch are double the picks per inch. Hence, while checking the repeat, keep the setting with the ends per inch doubled to the picks per inch. By this, the four repeats (2×2) of the weave on 32×16 size will result in the square size, as seen in Figure 8.4f. Furthermore, set the graph count as 16×8 proportionate to ends per inch and picks per inch, which makes the thick grid lines result in the square grid, as also seen in Figure 8.4f.

For creating the weft-backed weave, the 'divide horizontal section' and 'merge horizontal section' options are used. For example, consider that it is required to create a weft-backed weave on 8×16 by combining a weft face twill on 8×8 (1 up-7 down) and a warp face twill on 8×8 (7 up-1 down) in a 1:1 pick-way. To achieve this, first, take a graph of 8×16 as shown in Figure 8.4g. Click on the 'divide horizontal section' option by taking the horizontal section (y) as 2. By doing this, the odd picks in the graph are grouped as the first section of eight picks on the bottom side, and the even picks in the graph are grouped as the second section of eight picks on the top side. Next, mark a 1 up-7 down twill in the bottom section of eight picks, as shown by the red colour (up) in Figure 8.4h and mark a 7 up-1 down twill in the top section of eight picks, as shown by the green colour (up) in Figure 8.4h. Then, click the 'merge horizontal section' option. By clicking, the picks in two sections get merged in a 1:1 pick-way, resulting in a weft-backed weave as shown in Figure 8.4i.

After that, check the repeating of the weave by taking four repeats (2×2). In the weft-backed weave, mostly, the picks per inch are double the ends per inch. Hence, while checking the repeat, keep the setting with picks per inch doubled to the ends per inch. By this, the four repeats of the weave on 16×32 size will result in the square size, as seen in Figure 8.4j. Furthermore, set the graph count as 8×16 proportionate to ends per inch and pick per inch, which makes the thick grid lines result in a square grid, as also seen in Figure 8.4j.

8.3.2 DOUBLE-CLOTH WEAVE—SELF-STITCHED

For creating a double-cloth self-stitched weave, both the horizontal and vertical sections of the divide and merge options are used. In the typical double-cloth weave, the face and back ends are in 1:1 order, as are the face and back picks. The odd ends are face ends and the even ends are back ends. Similarly, odd picks are face picks and even picks are back picks. The five steps of creating the self-stitched double-cloth weave are: (i) face picks interlace with face ends as per the face weave; (ii) while interlacing the face picks, all the back ends are down; (iii) back picks interlace with back ends as per the back weave; (iv) while interlacing the back picks, all the face ends are up; these four steps form the two layers of cloth one above the other; and (v) stitching of double cloth is done either by lifting a few back ends while interlacing the face picks with face ends or by lowering a few face ends while interlacing back picks with back ends. This results in forming the stitched double cloth. From the above steps, it is clear that face picks and back picks have their interlacing with face and back ends in a defined order. Hence, it is easy to create a double-cloth weave first by grouping or sectioning the ends as face ends and back ends, as well as the picks as face picks and back picks. The sectioning is done by using both the horizontal and vertical sections of the divide and merge options. Figure 8.5a shows the sectioning of ends and picks and the first four steps of creating the double-cloth weave.

For creating the double-cloth weave, as an example, consider the face weave as 3 up 1 down twill (4 threads) and the back weaves as 2 up 2 down twill (4 threads).

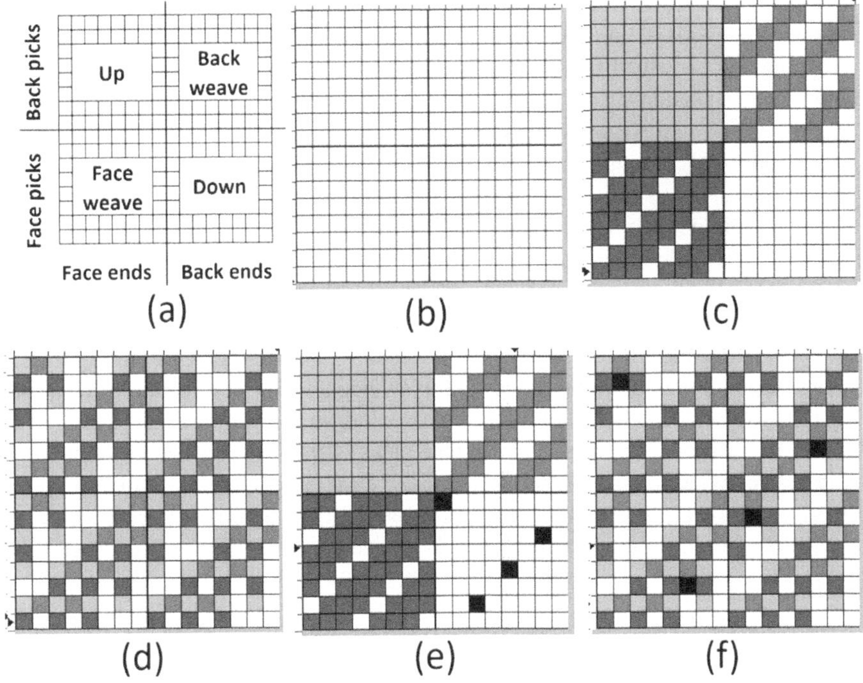

FIGURE 8.5 Stitched double-cloth weave creation using the divide and merge options of CAFGD: (a) the first four steps of creating double-cloth weave, (b) a 16 × 16 graph is taken, divided into two vertical sections of face ends and back ends and two horizontal sections of face picks and back picks, (c) face weave, back weave, ends down, and ends up being marked in the respective sections, (d) two layers of cloth created without stitching after merging of vertical and horizontal sections, (e) stitching marks are given, and (f) double-cloth weave is created with stitching after merging of vertical and horizontal sections.

For creating one repeat of the double-cloth weave, first, take 16 × 16 as shown in Figure 8.5b to have 2 repeats of the face and back weaves and the stitching repeating on 16 × 16. Next, click on the 'divide vertical section' option by taking the vertical section (x) as 2, and again, click on the 'divide horizontal section' option by taking the horizontal section (y) as 2. By doing this, the odd ends on the graph are grouped as the first section of eight ends at the bottom left side, and the even ends on the graph are grouped as the second section of eight ends at the right side. Furthermore, the odd picks in the graph are grouped as the first section of eight picks on the bottom side, and the even picks in the graph are grouped as the second section of eight picks on the top side as indicated in Figure 8.5a.

With this, (i) the 8 × 8 bottom-left-side section of the graph indicates the interlacing section of face picks with the face ends as per the face weave. Therefore, in this section, input the face weave (3 up-1 down twill), as shown by the red colour in Figure 8.5c. (ii) The 8 × 8 top-right-side section of the graph indicates the interlacing section of back picks with back ends as per the back weave. Therefore, in this section, input the back weave (2 up-2 down twill), as shown by the magenta colour in Figure 8.5c. (iii) The 8 × 8 bottom-right-side section of the graph indicates the lowering of the back ends for the face picks. Therefore, in this section, leave the yellow colour as it is, as shown in Figure 8.5c. (iv) The 8 × 8 top-left-side section of the graph indicates the lifting of the face ends for the back picks. Therefore, the blue colour in this section shows the lifting of all face ends. After completing the inputs as above, click on the 'merge vertical section' option and the 'merge horizontal section' option. By clicking, the ends in two sections get merged in 1:1 order, and the picks in two sections also get merged in 1:1 order as well, resulting in a double-cloth weave—without stitching—as shown in Figure 8.5d.

The weave created above will result in two layers of cloth one above the other without any stitching. To create the double cloth with stitching, first complete the four steps as described above. Then, additionally, it is necessary to give stitching points as described in the fifth step of forming the self-stitched double cloth. Figure 8.5e shows the method of giving stitching points by lifting a few back ends while interlacing the face picks, as indicated by the black colour in the bottom right section of the graph. The lifting of the back ends on face picks is chosen in such a way that the lift is in between the float of the face ends on the face layer. Only four stitching points are given for two repeats of the weave. After giving the stitching, click on the 'merge vertical section' option and the 'merge horizontal section' option. By clicking, the ends in two sections get merged in 1:1 order, and the picks in two sections also get merged in 1:1 order as well, resulting in a double-cloth weave with stitching as shown in Figure 8.5f.

8.4   FLOAT AND FLOAT CONTROL IN THE FIGURED GRAPH

8.4.1   TYPES OF FLOATS AND FLOAT LENGTH

Float in the fabric is commonly defined as 'one series of yarn going down or coming up for more than one thread of another series'. End float in the fabric is defined as 'an end is going down or coming up for more than one pick'. A pick float in the fabric is defined as 'a pick going down or coming up for more than one end'. Plain weave is formed with 1 up, 1 down, and 1 down, 1 up. Hence, there is no float in the plain weave. All the other weaves are formed by floats (that is, more than 1 up and 1 down). The floats that occur in the fabric are of four types. Figure 8.6a shows these four types of floats formed in different places of a 10 × 10 honeycomb weave marked on 20 × 20 (2 × 2 repeats). In this figure, blue represents the end up, and yellow represents the end down (pick up). Table 8.1 describes these four floats, the place they occur on the cloth, and their identification from the honeycomb weave shown in Figure 8.6a.

The float length of the end is measured by the number of picks over or below which the end is floating. In the vertical cross-section 1, shown on the right side of Figure 8.6a, the end is floating over the 7 picks. Hence, 7 is the float length of the first end floating on the face side. In the vertical cross-section 2, shown on the left side of Figure 8.6a, the end is floating below the 9 picks. Hence, 9 is the float length of the sixth end floating down at the backside. The float length of the pick is measured by the number of ends over or below which the pick is floating. In the horizontal cross-section 3, shown at the bottom side of Figure 8.6a, the pick is floating above the 9 ends. Hence, 9 is the float length of the first pick floating on the face side. In the horizontal cross-section 4, shown on the top side of Figure 8.6a, the pick is floating below the 7 ends. Hence, 7 is the float length of the sixth pick floating down at the backside.

The maximum length of the float allowed in the figured fabric is decided in such a way that the floats are not liable to slip or fray when the fabric is subjected to strain and friction during wear. The maximum float length depends upon the threads per unit space of the figured fabric. Normally, the maximum float length is taken as 'one-tenth of the threads per inch'. For example, consider that a fabric is woven with 90 ends per inch and picks per inch (EPI and PPI). One-tenth of 90 is 9. Therefore, in this fabric, the maximum float length allowed is 9 or below 9. The end float length of

---

TABLE 8.1

Description of Four Different Floats, the Place They Occur on the Cloth, and Their Identification from the Graph

Description of the Float	Float Occurring Side While Weaving	Float Identification on the Graph Given in Figure 8.6a	Representation of Cross-Section of Float in Figure 8.6a
Vertical End Float			
End floating over the picks	on the face side	Continuous vertical marks (1)	1—on the right side of Figure 8.6a
End floating below the picks	on the back side	Continuous vertical blanks (2)	2—on the left side of Figure 8.6a
Horizontal Pick Float			
Pick floating over the ends	on the face side	Continuous horizontal blanks (3)	3—at the bottom of Figure 8.6a
Pick floating down the ends	on the back side	Continuous horizontal marks (4)	4—on the top of Figure 8.6a

more than 9 has to be controlled/bound by the pick binding. Similarly, the pick float length of more than 9 has to be controlled/bound by the end binding.

8.4.2 Float Identification and Float Control

In the CAFGD software, there are two options for identifying the floats. The first one is the 'identify vertical end float' option. Using this option, it is possible to find the end floats of the required length. The second one is the 'identify horizontal pick float' option. Using this option, it is possible to find the pick floats of the required width. A figured graph of 40 × 40 is shown in Figure 8.6b. In this graph, the centre circle in red colour is warp up. The green colour ring after the centre red circle is weft up. The red colour ring after the green ring is warp up. The outer ground in yellow colour is weft up.

First, the maximum end and pick floats allowed in this figured graph are decided. It is decided as 9, considering that the ends and picks per inch are 90. In the inner red circle, 1 down-7 up twill weave is mapped as shown in Figure 8.6b. Hence, the float length in this circle is less than 9. Next, to identify the horizontal pick floats in the outer red ring (warp), the maximum pick float length is fixed at 9. The figure colour (red) in which the floats have to be identified is taken as the background colour, and any new colour (say, blue) is taken as the foreground colour. Then, click on the 'identify horizontal pick' option. By clicking, the blue colour taken in the foreground gets applied horizontally over the red ring and inner red circle, wherever the horizontal pick floats are less than 9, leaving the horizontal pick floats in red colour, which are more than 9, as shown in Figure 8.6c. The length of all the horizontal pick floats thus identified in red is more than 9. Now, suitable binding marks are applied to control these picks floating on the back side. The yellow colour, which is a weft colour, is used for binding the red floats. The binding weave taken is 8 thread sateen with 3 moves. The binding is applied manually over the red colour. After completion of giving binding in all the places, the blue colour is changed again into red colour. The graph at this stage is shown in Figure 8.6d where the bottom and top weft floats of the outer red ring are stitched along with the binding marks on the inner circle.

After that, to identify the vertical end floats on the outer red ring (warp), the maximum end float length is fixed at 9. The figure colour (red) in which the floats have to be identified is taken as the background colour, and any new colour (say, blue) is taken as the foreground colour. Then, click on the 'identify vertical end float' option. By clicking, the blue colour taken in the foreground gets applied vertically over the red ring and inner red circle, wherever the vertical end floats are less than 9, leaving the vertical end floats in red colour, which are more than 9, as shown in Figure 8.6e. The length of all the vertical end floats thus identified in red is more than 9. Now, suitable binding marks are applied to control these red end floats floating on the face side. The yellow colour, which is a weft colour, is used for binding the red floats. The binding weave taken is 8 thread sateen with 3 moves. The binding is given manually over the red colour. After completion of applying binding in all the places, the blue colour is changed again to red colour. The graph at this stage is shown in Figure 8.6f. In these two steps, both the end and pick floats in the outer red ring are identified and given suitable binding.

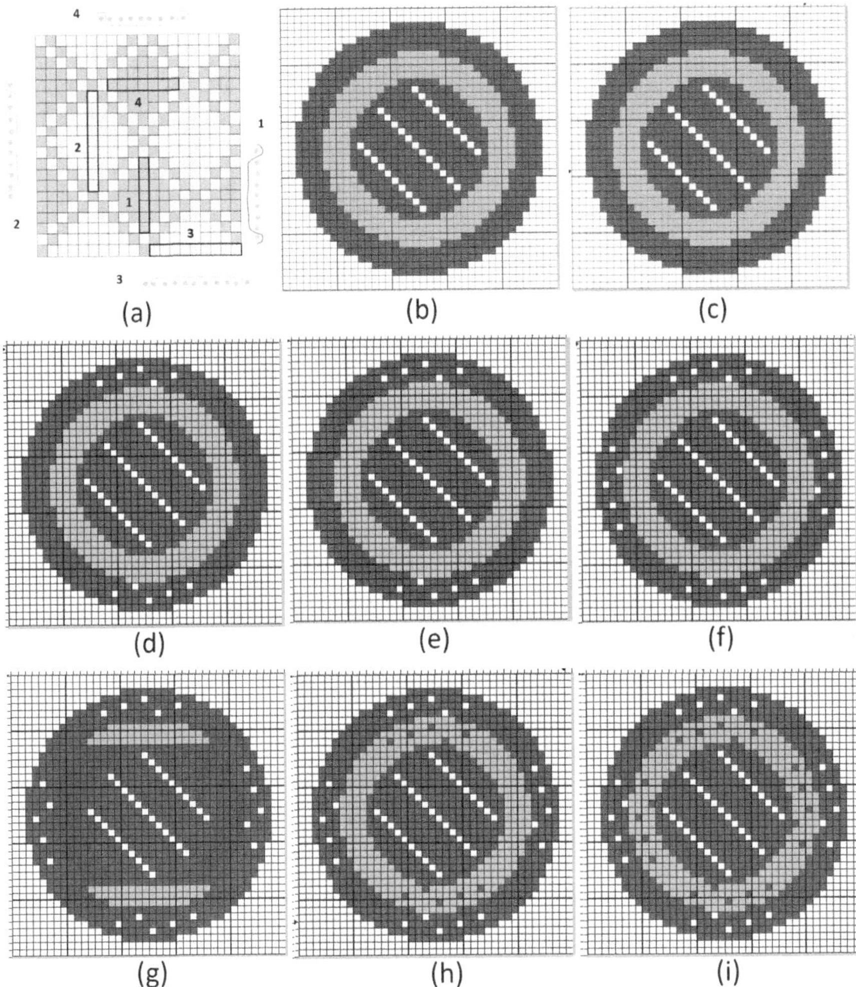

FIGURE 8.6 Graphs showing the float, float length, float identification, and float control: (a) four types of floats shown in honeycomb weave, (b) a figured graph on 40 × 40, (c) horizontal pick floats are identified in the red ring, (d) sateen binding weave is given for the horizontal floats identified, (e) vertical end floats are identified in the red ring, (f) sateen binding weave is given for the vertical floats identified, (g) horizontal pick floats are identified in the green ring, (h) sateen binding weave is given for the horizontal floats identified, and (i) final graph completed after binding the vertical end floats.

Next, to identify the horizontal pick floats in the inner green ring (ground-weft), the maximum pick float length is fixed at 9. The ground colour (green) in which the floats have to be identified is taken as the background colour, and any new colour (say, blue) is taken as the foreground colour. Then, click on the 'identifying horizontal pick float' option. By clicking, the blue colour taken in the foreground gets applied horizontally over the green ring, wherever the horizontal pick floats are less than 9, leaving the horizontal pick floats in green colour, which are more than 9, as shown in Figure 8.6g. The length of all the horizontal pick floats thus identified in green is more than 9. Now, suitable binding marks are applied to control these picks floating on the back side. The red colour, which is a warp colour, is used for binding the green floats. The binding weave taken is 8 thread sateen with 3 moves. The binding is applied manually over the green colour. After completion of giving binding in all the places, the blue colour is changed again into the green colour. The graph at this stage is shown in Figure 8.6h where the bottom and top weft floats of the inner green ring are stitched. Then, the vertical floats in the green inner ring (weft) are identified and stitched by following the procedure described above. The final graph, thus completed after binding the floats in all the figure parts, is shown in Figure 8.6i. Of course, the ground yellow weft should also be given suitable binding by identifying the floats.

BIBLIOGRAPHY

1. ArahPaint. (2018). *User's Manual*. Arahne Software Company in Ljubljana, Slovenia.
2. Auto Tex 2000. (2005). *User's Manual*. PLC Consulting Company, New Delhi.
3. CadVantage Win Jacquard. (2001). *User's Manual*. Teckmen Systems Pvt. Ltd., Pune.
4. Grosicki, Z. J. (2004). Figuring with extra threads. In *Watson's Advanced Textile Design* (pp. 11–41). Woodhead Publishing Limited, Cambridge.
5. Textronics. (2000). *User's Manual of Design*. Jacquard Textronics Design Systems (I) Pvt. Ltd., Mumbai.

9 Graph Preparation for Different Figured Fabric Varieties

9.1 UNCONTROLLED AND CONTROLLED WEAVE MAPPING

In the manual figured graph designing, while applying the binding weave marks, the designer restricts/controls where to give and not to give binding marks as per the graph variety produced and float length allowed, whereas in the Computer-Aided Figured Graph Designing (CAFGD), sequences of steps are followed to avoid the binding weave marks in the required boundary of the motif.

In both manual and CAFGD, the conditions to be born in mind while applying different weaves on different figure parts and ground are:

i. The weave marks are applied as per the variety of cloth structures to be produced,
ii. The weave marks applied should enhance the appearance of the figure part distinctly from the ground,
iii. The weave marks applied should control long floats. It is avoided where there is no longer float, and
iv. The weave marks applied should not distort the boundary of the figure and ground that was edited as per the outline of the given motif.

Giving or avoiding binding weave marks on the one or two pixels of top, bottom, left, right, and corners of the ground and figure boundary depends upon the warp and weft floats formed in the different figured varieties. This is understood by taking four different varieties of jacquard graph, viz. double cloth, extra warp, extra weft, and single cloth (Damask) as first examples. These varieties are taken because they cover the application of equal face, warp face, weft face, and both warp-weft face weaves. Moreover, these varieties are separately or combined and used in the textile industry to produce many figured fabrics such as bedsheets, home furnishings, and traditional and contemporary jacquard sarees. In sarees, the borders are woven in the extra warp or extra weft principle. The extra weft figuring principle is used for the pallu part. The extra warp or extra weft figuring principle is used for buttas, and the single-cloth (damask) principle is used in the body part of the saree.

The different weaves applied in the weave mapping process are in different colours. Hence, after mapping, all the colours that indicate end up in all these weaves

DOI: 10.1201/9781003441205-9

are converted into warp colours used in the fabric. Similarly, all the colours that indicate end down in all these weaves are converted into weft colours.

9.2 UNCONTROLLED MAPPING OF EQUAL FACE WEAVE— INTERCHANGING PLAIN DOUBLE CLOTH

In the figured graph designing, equal face weaves such as plain, rib, mat, two up two down twill, and interchanging plain double cloth (IPDC) are used. They interlace warp and weft colours equally. These weaves contain minimal–equal/unequal warp and weft floats. *These weaves are applied (mapped) throughout the figure and ground without restricting it to the corner pixels of the boundary.* This is called the 'uncontrolled application of weave marks' on IPDC. Graph preparation for IPDC using equal face weaves is one of the examples. This variety is woven with one series of warp and weft. The warp is in two colours (light and dark) in the 1:1 ends order. The weft is also in two colours (light and dark) in the 1:1 picks order. Table 9.1 gives the details of the edited graph design used for preparing the figured graph for IPDC and the two equal face weaves used for the same.

A repeat of the final edited jacquard graph design on 48 ends × 48 picks is shown in Figure 9.1a. The weave for producing light-coloured plain double cloth in the ground (on the face side) is shown in Figure 9.1b, and dark-coloured plain double cloth in the figure (on the face side) is shown in Figure 9.1c. These two IPDC weaves are opposite to each other. In Figure 9.1d, the ground weave (Figure 9.1b) is mapped on the ground colour (yellow) in the edited graph. In Figure 9.1e, the figure weave (Figure 9.1c) is also mapped on the figure colour (red) on the edited graph. It is the final graph of the figured IPDC. This graph is without any long end and pick floats and with no distortion of the figure and ground boundary. The flowchart given in Figure 9.2 shows the sequence of steps followed in preparing the figured graph for IPDC.

TABLE 9.1

Description of the Edited Graph, Two Weaves of Interchanging Plain Double Cloth, and Colours Used

Figure No.	Figure Description	Colour	Colour Description
Figure 9.1a	Edited graph; Interchanging plain double cloth	Colour – A (Yellow)	Ground –Light + Light (LE + LP)
		Colour – B (Red)	Figure – Dark + Dark (DE + DP)
Figure 9.1b	Ground weave-1 graph on 4 × 4; (LE + LP) – Face; (DE + DP) – Back	Colour – A (Yellow) Colour – C (Blue)	Pick up End up
Figure 9.1c	Figure weave-2 graph on 4 × 4; (DE + DP) – Face; (LE + LP) – Back	Colour – B (Red) Colour – D (Green)	End up Pick up

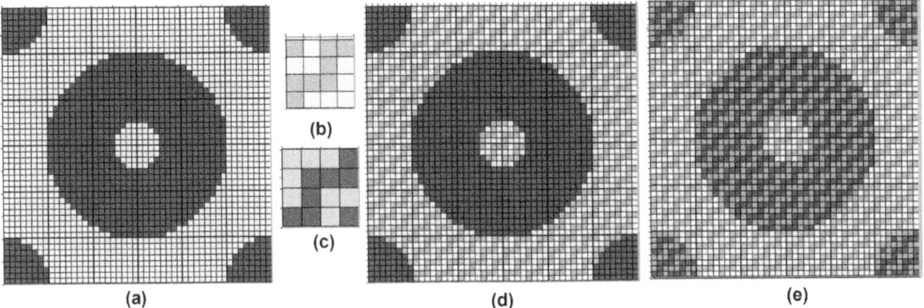

FIGURE 9.1 Preparing the figured interchanging plain double cloth graph: (a) edited graph design on 48 × 48, (b) ground weave, (c) figure weave, (d) mapping the ground weave with the ground colour, and (e) mapping the figure weave with the figure colour; final figured interchanging plain double cloth graph.

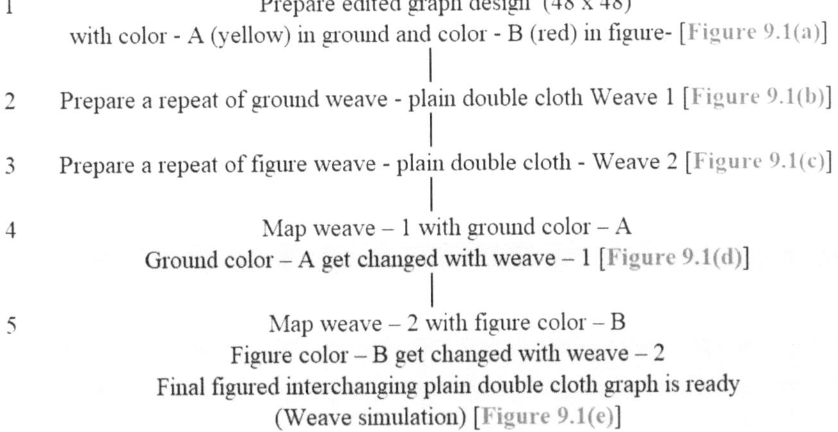

1   Prepare edited graph design (48 x 48)
with color - A (yellow) in ground and color - B (red) in figure- [Figure 9.1(a)]

2   Prepare a repeat of ground weave - plain double cloth Weave 1 [Figure 9.1(b)]

3   Prepare a repeat of figure weave - plain double cloth - Weave 2 [Figure 9.1(c)]

4   Map weave – 1 with ground color – A
Ground color – A get changed with weave – 1 [Figure 9.1(d)]

5   Map weave – 2 with figure color – B
Figure color – B get changed with weave – 2
Final figured interchanging plain double cloth graph is ready
(Weave simulation) [Figure 9.1(e)]

FIGURE 9.2 Flowchart showing the algorithm to prepare the graph for figured interchanging plain double cloth.

9.3 UNCONTROLLED MAPPING OF WARP AND WEFT FACE WEAVES

In the figured fabric, the figure and ground are formed by warp float and weft float or both by warp and weft float. It is essential to control the horizontal weft floats by the warp binding (weft face weave with warp binding). Similarly, the vertical warp floats must be controlled by the weft binding (warp face weave with weft binding).

It is essential to understand how the boundary of the edited graph is distorted when the warp and weft face binding weaves are applied without restricting it in the boundary. The example is shown in Figure 9.3. Figure 9.3a shows the eight-thread weft face (sateen) binding weave selected for the ground. In this weave, yellow is end down and blue is end up. Figure 9.3b shows the eight-thread warp face (satin) binding weave selected for the figure. In this weave, green indicates end down, and red

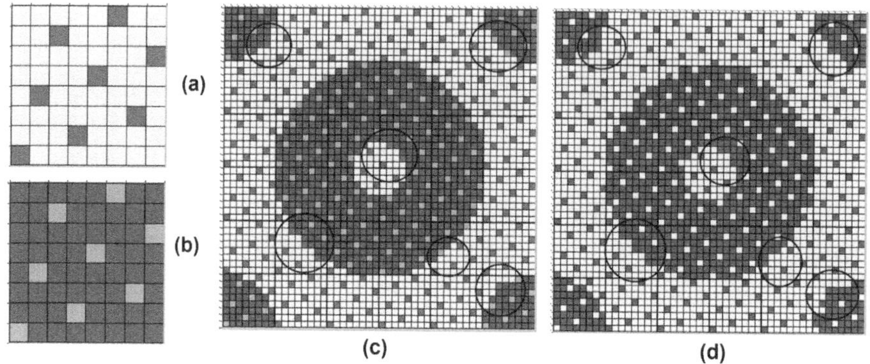

FIGURE 9.3 Uncontrolled application of warp and weft face weaves: (a) sateen binding weave −1 on 8 × 8 for ground, (b) satin binding weave −2 on 8 × 8 for figure, (c) mapping the ground and figure weaves, and (d) distortion of ground and figure boundary after weave application.

indicates end up. Figure 9.3c shows the mapping of these two weaves for binding the floats on the ground and the figure parts of the edited graph (Figure 9.1a). The weave is applied without any control/restriction. It can be observed in Figure 9.3c that the blue binding marks on the yellow ground part touch the corners of the red-figure boundary in many places. Similarly, green binding marks on the red-figure parts touch the corners of the yellow ground boundary in many places. These are shown by the black circles. To convert the graph into two colours (yellow for end down and red for end up), the end up in the ground (Blue) is converted to red and the end down in the figure (green) is converted to yellow as seen in Figure 9.3d. In this graph, the distortion of the edited boundary can be seen clearly in many places which are shown by the black colour circles.

9.4 CONTROLLED MAPPING OF WEAVES USING COLOUR CONTOUR OPTIONS

From Figure 9.3, it could be understood that controlled/restricted weave application is a must to avoid the binding marks of warp face and weft face weaves in the non-required places to retain the correct edited boundary of the figure and ground parts. It is essential to follow two points for getting the perfect edited boundary of the warp and weft face figure and ground. They are (i) the binding marks are not given in the one or two corner pixels of the boundary and (ii) the binding marks are mapped if there is a long float other than the one or two pixels of the corners of the boundary. In CAFGD, the figure and ground areas are applied with different colours according to the different weaves to be mapped (one colour = one weave). After this, other colours are applied in the boundary of the ground and figure areas where the binding marks are to be restricted. Giving separate colours in one or two pixels of the ground and figure boundary is called 'Contouring'.

Presently, in most of the CAFGD software, two methods of contouring are used for the float control application of weave marks. These are (i) contouring using the 'All-directional combined colour contour option' and (ii) contouring using the 'Selective directional colour contour option'. As the name implies, in the all-directional combined colour contour option, one or two pixels in *all directions* of the boundary (complete boundary) *are combined* and given a colour. In the selective directional colour contour option, one or two pixels on the *selected directions* of the boundary *are separated* and given a colour. Some software has only all-directional contour options and some have both.

Figure 9.4 shows different colour contour options. Figure 9.4a shows the right side (half) of the edited graph shown in Figure 9.1a. While using the contouring option, if all directions are selected as shown in Figure 9.4b, the complete outline is given in the boundary of the edited graph. The outline is for one or two pixels in all eight directions combined (top, top right, right, right bottom, bottom, bottom left, left, and left top) as shown in Figure 9.4c. While using the contouring option, if a particular direction/directions are selected, it is called 'Selective directional colour contouring'. By this, the border outline is given for one or two pixels in the selective direction/directions, viz. top, bottom, left, and right directions.

In the figured extra warp, there are two series of warps, viz. ground and figuring warp, and only one series of the weft. Hence, in the figured extra warp graphing, there are only vertical extra warp floats to be controlled. There is no weft float because the weft is interwoven in plain order with the ground warp. Hence, the binding marks are given only by observing the long vertical end floats (extra warp). The binding marks are avoided at the one or two pixels of the top and bottom edges in each vertical end float. For doing this, top and bottom directions are selected while using the contouring option, as shown in Figure 9.4d. Figure 9.4e shows the identifying (contouring) of two pixels on the top and bottom edges, in the boundary of the edited graph given in Figure 9.4a.

In the figured extra weft, there are two series of wefts, viz. ground and figuring weft, and only one series of warp. Hence, in the figured extra-weft graphing, there are only horizontal extra weft floats to be controlled. There is no warp float because the warp is interwoven in plain order with ground weft. Therefore, the binding marks are given only by observing the long horizontal weft floats (extra weft). The binding marks are avoided at the one or two pixels of left and right edges in each horizontal float. For doing this, left and right directions are selected while using the contouring option, as shown in Figure 9.4f. Figure 9.4g shows the contouring of two pixels at the left and right edges, in the boundary of the edited graph given in Figure 9.4a.

In the figured single cloth (damask), there is only one series of warp and weft. Warp forms the figure; the weft forms the ground. Hence, in figured single-cloth graphing, there are vertical warp floats and horizontal weft floats to control. Therefore, the binding weave marks are given by observing both long vertical and horizontal floats. The binding weave marks are restricted only to one or two corner pixels. *In most of the software, for identifying the corner pixels, if corner directions are selected while using the contouring option as shown in Figure 9.4h, it gives a complete outline of the boundary of the edited graph. A new procedure is derived to identify the corner pixels in the boundary of the edited graph using the layer merging option and transparent colour option. Figure 9.4i shows the contouring of only two corner pixels, in the boundary of the edited graph shown in Figure 9.4a using the new procedure.*

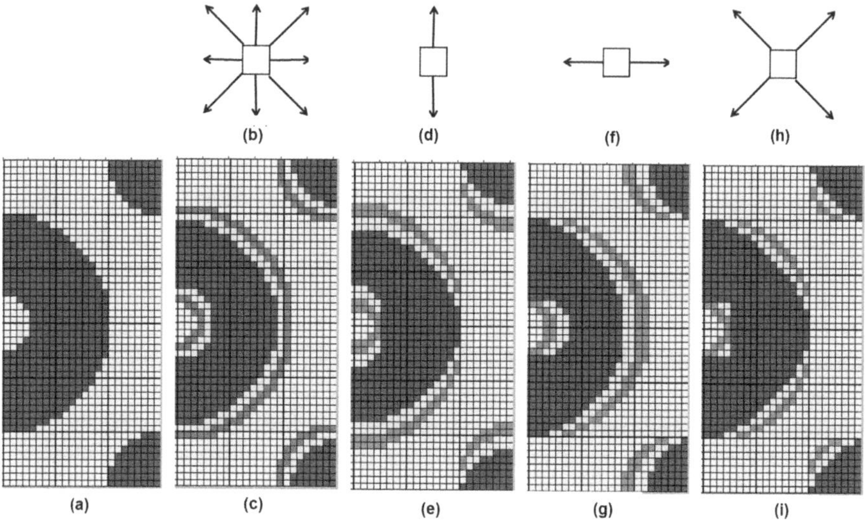

FIGURE 9.4   Different contour options for the float control weave application: (a) a part of edited graph, (b) all-directional combined contour option, (c) contouring the edited graph in all directions, (d) top and bottom—selective directional contour option, (e) contouring the edited graph in the top and bottom directions, (f) left and right—selective directional contour option, (g) contouring the edited graph in left and right directions, (h) corners – selective directional contour option, and (i) contouring the edited graph only in the corners.

Table 9.2 describes the edited graphs (Figure 9.1a and 9.4a) and the binding weaves (Figure 9.3a and b). In this Table, the colours in the edited graph and the binding weaves are also described for extra warp, extra weft, and single cloth for comparative understanding. Table 9.3 compares the directions in which the float is controlled, the pixels where the binding marks are to be given and not to be given for the figured fabrics of IPDC, extra warp, extra weft, and single cloth.

9.5   PREPARING THE FIGURED GRAPHS USING THE ALL-DIRECTIONAL COLOUR CONTOUR OPTION

There are three steps in preparing the figured graph using the 'all-directional combined contour option'. Out of these, the first two steps are common for preparing the figured extra warp, extra weft, and single-cloth graphs with warp and weft face weave. They are (i) contouring the boundary of ground and figure areas using all-directional contour options irrespective of any figured fabric; (ii) applying the binding weave other than the contour places; that are inside the figure and ground area. With these two steps, a figured weave graph is prepared with floats for float checking. The third step is to identify the floats in the weave graph and apply the binding for the floats in the contour. This third step varies as per the variety of figured fabric for which the graph is prepared, viz. extra warp, extra weft, and single cloth.

The flowchart shown in Figure 9.5 shows the common algorithm followed in preparing the jacquard weave graph for float checking as described above. Figure 9.6a

TABLE 9.2

Description of the Edited Graph, Two Weaves, and Colours for Different Figured Fabrics

Figure No. & Description	Colours	Colours Description		
		Extra Warp	**Extra Weft**	**Single Cloth**
Figure 9.1a Figure 9.4a Edited graph	Ground Colour – A (Yellow)	Extra warp down	Extra weft up	Weft up
	Figure Colour – B (Red)	Extra warp up (Extra warp figure formed on the face side while weaving)	Extra weft down (Extra weft figure formed at the back side while weaving)	Warp up
Figure 9.3a Sateen binding on 8 × 8 for ground (weave −1)	Ground Colour – A (Yellow)	Extra warp down in the ground	Extra weft up in the ground	Weft up in the ground
	Binding Colour – C (Blue)	Extra warp up binding in 8 × 8 sateen order	Extra weft down binding in 8 × 8 sateen order	Warp up binding in 8 × 8 sateen order
Figure 9.3b Satin binding on 8 × 8 for the figure (weave – 2)	Figure Colour – B (Red)	Extra warp up in figure	Extra weft down in figure	Warp up in figure
	Binding Colour – D (Green)	Extra warp down binding in 8 × 8 sateen order	Extra weft up binding in 8 × 8 sateen order	Weft up binding in 8 × 8 sateen order

shows the graph where contouring is applied for the complete boundary in the ground and the figure colours of the edited graph are shown in Figure 9.6a. Figure 9.6b shows the graph in which the ground and figure weaves are mapped with the ground and figure colours, respectively. Figure 9.6c shows the graph in which the ground and figure contour colours are again converted to the ground and figure colours, respectively. It can be seen in the graph that the long floats remain at the top, bottom, left, and right edges of the contouring on the figure and ground. The long floats are indicated by the black colour ellipses. This graph is a 'weave graph for float checking' or 'weave graph before float checking' or 'weave graph with floats' and taken further for float checking as per the figure variety.

9.5.1 Extra Warp-Figured Graph

The left side of the flowchart shown in Figure 9.7 gives the sequence of steps followed to prepare the final graph for the figured extra warp and the right side shows the figured extra weft. For preparing the graph for the figured extra warp, it is necessary to identify only the vertical end floats in the ground and the figure of the weave graph with floats (Figure 9.6c). Figure 9.8a shows the graph in which the vertical end

TABLE 9.3

Comparison of Float Control Direction and Binding Marks Control for Different Figured Fabrics

Direction of Pixels and Reference Figure No.	Interchanging Plain Double Cloth	Extra Warp	Extra Weft	Single Cloth
Weave	Minimum equal warp and weft float	Only warp float	Only weft float	Both warp and weft float
The direction in which the float is controlled	No floats	Only vertical end float is controlled	Only horizontal pick float is controlled	Both vertical end float and horizontal pick float are controlled
1 or 2 pixels of top & bottom edge of stepping (Figure 9.4d and e)		**Binding marks are not given**	Binding marks are given	Binding marks are given
1 or 2 pixels of left & right edge of stepping (Figure 9.4f and g)		Binding marks are given	**Binding marks are not given**	Binding marks are given
1 or 2 pixels of the corner of the stepping (Figure 9.4h and i)		**Binding marks are not given**	**Binding marks are not given**	**Binding marks are not given**

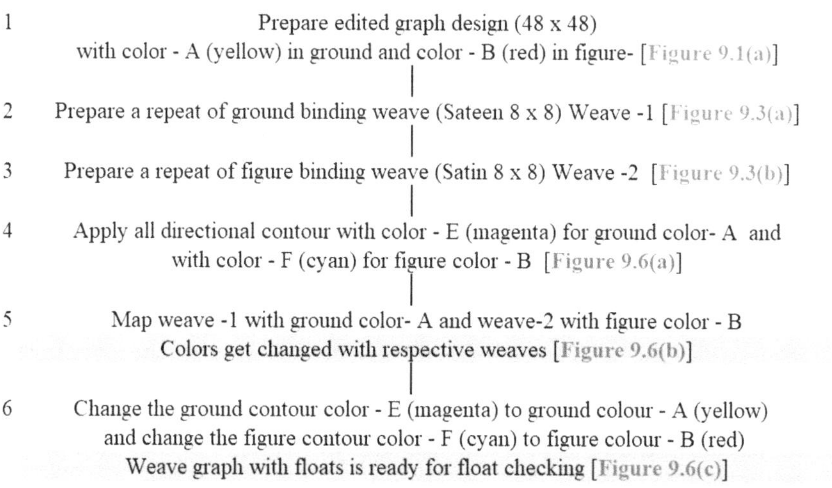

FIGURE 9.5   Flowchart showing the common algorithm to prepare the figured weave graph with warp and weft face weaves using all-directional contour options.

floats of more than 10 pixels in length in the ground are identified and given binding. Similarly, Figure 9.8b shows the identification and binding of vertical end floats in the figure. Now, the magenta colour taken for identifying the floats in yellow colour is changed to yellow colour. Similarly, the cyan colour taken for identifying the floats

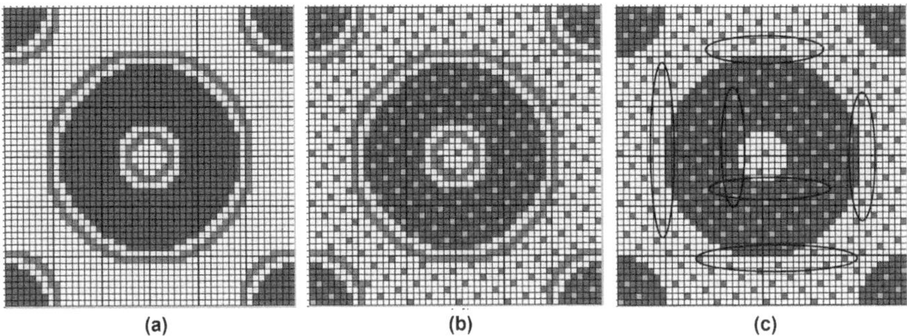

(a)   (b)   (c)

FIGURE 9.6 Preparing the figured weave graph for float checking: (a) contouring the ground and figure colours, (b) mapping the ground and figure weaves, and (c) figured weave graph with length y floats to be checked.

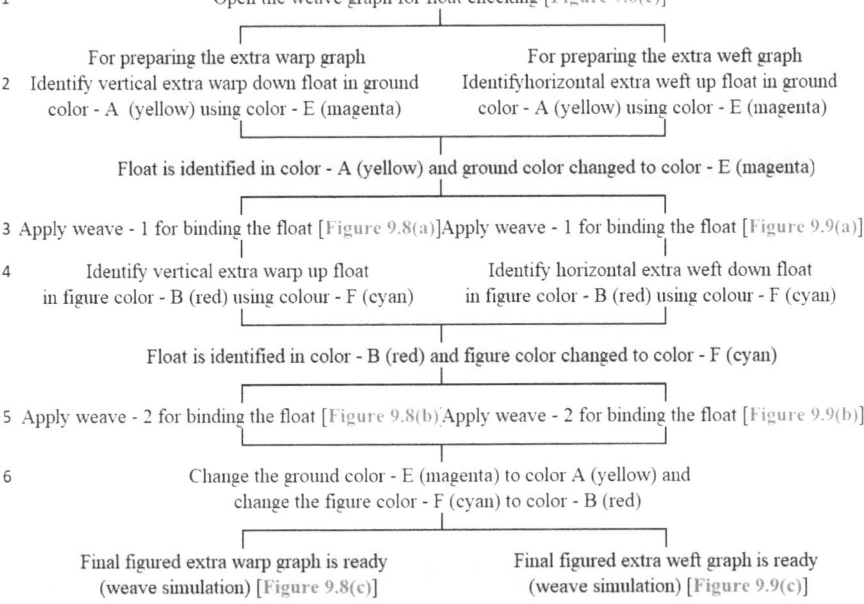

FIGURE 9.7 Flowchart showing the algorithm to prepare the final graph for figured extra warp and extra weft from the weave graph with floats.

in red colour is changed to red colour. With this, the final jacquard graph for the figured extra warp is ready. This graph is without any long vertical end floats both in the figure and ground. It is without any binding marks on the top and bottom edges. Hence, the boundaries of the ground and figure are retained without any distortion as shown in Figure 9.8c. In this process, it is not necessary to identify and control the horizontal pick floats.

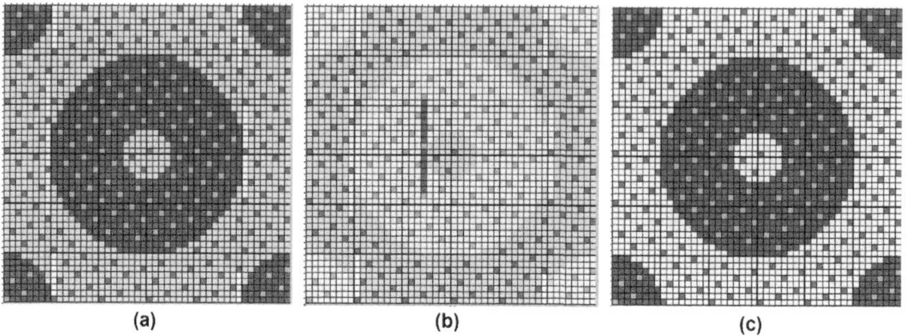

FIGURE 9.8   Preparing the figured graph for extra warp: (a) opening the weave graph with lengthy floats (Figure 9.6c), identifying lengthy vertical end floats in the ground, and binding them, (b) identifying lengthy vertical end floats in the figure and binding them, and (c) lengthy float-checked final figured graph for extra warp.

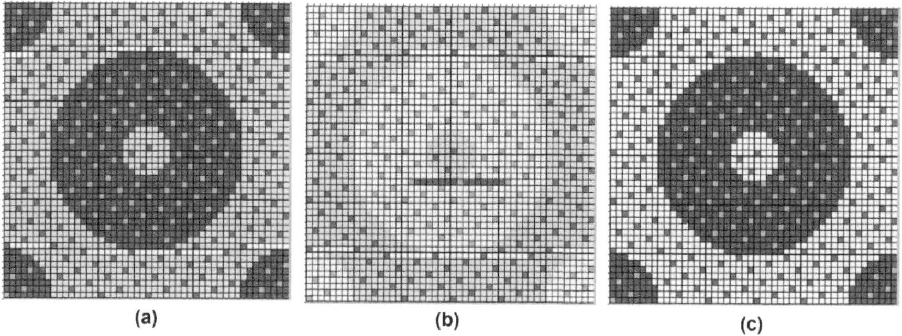

FIGURE 9.9   Preparing the figured graph for extra weft: (a) opening the weave graph with lengthy floats (Figure 9.6c), identifying lengthy horizontal pick floats in the ground and binding them, (b) identifying lengthy horizontal pick floats in the figure and binding them, and (c) lengthy float-checked final figured graph for extra weft.

9.5.2   Extra Weft-Figured Graph

For preparing the graph for figured extra weft, it is necessary to identify only the horizontal pick floats in the ground and the figure of the weave graph with floats (Figure 9.6c), instead of identifying vertical end floats as done in the figured extra warp graph preparation. Hence, the procedure to prepare the figured extra-weft graph narrated on the right side of the flowchart (Figure 9.7) is the same as the procedure followed to prepare the figured extra warp graph described on the left side of the flowchart (Figure 9.7), with only change in steps 2 and 4. That is, in steps 2 and 4, vertical end floats are identified for the extra warp, whereas horizontal pick floats are identified for the extra weft.

Figure 9.9a shows the graph in which the horizontal pick floats of more than 10 pixels in length in the ground are identified and given binding. Similarly, Figure 9.9b shows the identification and binding of horizontal pick floats in the figure as described above. Now, the magenta colour taken for identifying the floats in yellow colour is changed to yellow colour. Similarly, the cyan colour taken for identifying the floats in red colour is changed to red colour. With this, the final jacquard graph for the figured extra weft is ready. This graph is without any long horizontal pick floats. It is without any binding marks at the left and right edges. Hence, the boundaries of ground and figure are retained without any distortion as shown in Figure 9.9c. In this process, it is not necessary to identify and control the vertical end floats.

9.5.3 Single Cloth-Figured Graph

In the single cloth, both the warp and weft floats are to be controlled. The final figured extra warp graph (Figure 9.8c) is without any vertical end floats. But in this graph, there are horizontal pick floats. Hence, in this graph, if the horizontal pick floats are identified and given binding, it becomes the final graph for the figured single cloth. Therefore, the preparation of a figured single cloth graph contains three phases. They are (i) preparing the figured weave graph for float checking as per the procedure shown in the flowchart given in Figure 9.5; (ii) preparing the figured extra warp graph using the weave graph (prepared in phase one) as per the procedure shown on the left side of the flowchart given in Figure 9.7; and (iii) preparing the figured single cloth graph using the extra warp graph (prepared in phase two) as per the procedure shown on the right side of the flowchart given in Figure 9.7 for preparing the extra-weft graph.

Hence, for preparing the figured graph for the single cloth, open the final figured graph prepared for the extra warp (Figure 9.8c) in which the vertical end floats are checked and controlled. Figure 9.10a shows the identification of long horizontal pick floats in the ground of the final extra warp graph. The horizontal pick floats are also given binding. After this, the long horizontal pick floats in the figures are also identified and given binding as shown in Figure 9.10b. Then, the magenta and cyan colours used to identify the floats are changed to yellow and red, respectively. With this, the final jacquard graph for the figured single cloth is ready. This graph is without any

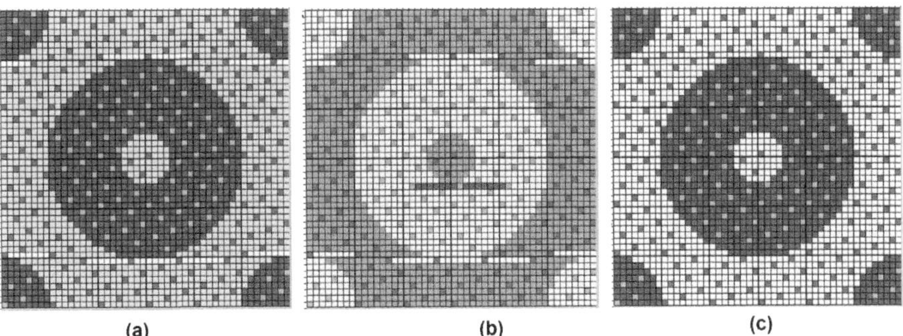

(a)   (b)   (c)

FIGURE 9.10 Preparing the figured graph for single cloth: (a) opening the lengthy vertical end floats checked extra warp graph Figure 9.8c, identifying lengthy horizontal pick floats in the ground, and binding them, (b) identifying lengthy horizontal pick floats in the figure, and binding them, and (c) lengthy float-checked final figured graph for single cloth.

long vertical end floats and horizontal pick floats. It is without any binding marks in the stepping corners. Hence, the boundaries of ground and figure are retained without any distortion as shown in Figure 9.10c.

9.6 PREPARING THE FIGURED GRAPHS USING THE SELECTIVE DIRECTIONAL COLOUR CONTOUR OPTION

While using the 'selective directional colour contour option', two steps are followed for preparing the figured extra warp, extra weft, and single-cloth graphs, with warp and weft face weaves. They are (i) contouring the boundary of the figure and ground area using the selective directional contour option. Selective contouring is applied as per the variety of figured fabric for which the graph is meant, viz. extra warp, extra weft, and single cloth. (ii) Applying the binding weave on places other than the contouring. That is, inside the figure and ground area.

9.6.1 Extra Warp-Figured Graph

The algorithm to prepare the extra warp-figured graph using the selective directional contour option is illustrated in the flowchart shown in Figure 9.11. For preparing the figured extra warp graph using the selective directional contour option, the edited graph is taken on 40 × 40 (Figure 9.12a). In the edited graph, the ground is in yellow. The ground weave is five threads of sateen (Figure 9.12b). The figure parts are given

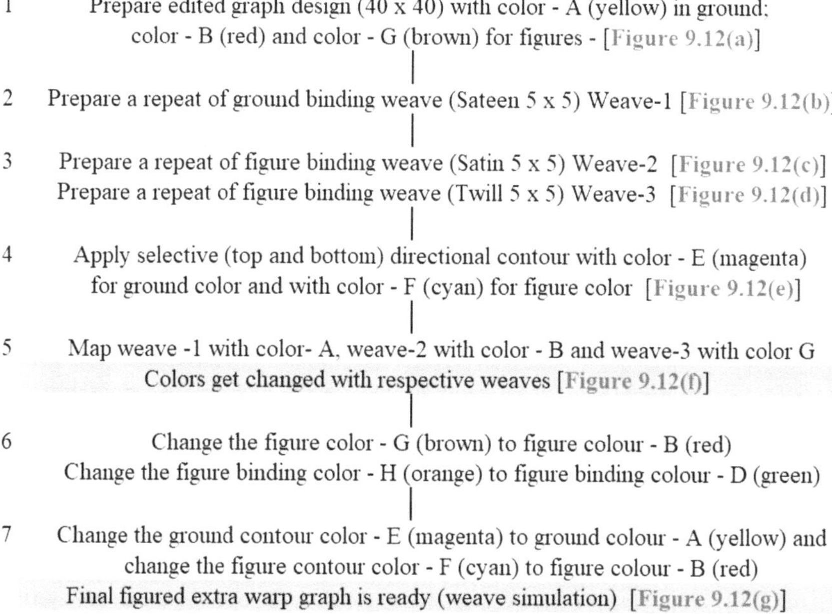

FIGURE 9.11 Flowchart showing the algorithms to prepare figured graph for extra warp using selective directional contour option.

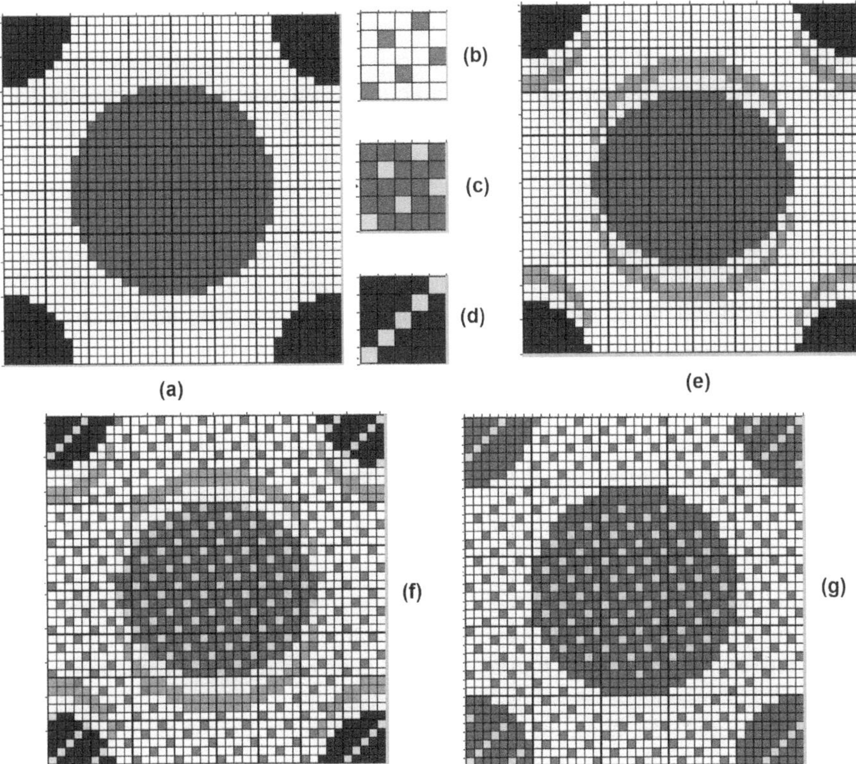

FIGURE 9.12 Preparing the figured extra warp graph using selective contour option (top and bottom): (a) edited graph on 40 × 40, (b) 5 threads sateen weave for the ground, (c) 5 threads satin weave for the figure, (d) 5 threads twill (4 up 1 down) weave for the figure, (e) top and bottom contouring in the ground and figure, (f) mapping ground and figure weaves, and (g) final float-controlled figured extra warp graph.

two colours (red and brown). The weave for red is five threads of satin (Figure 9.12c). The weave for brown is five threads of warp face twill (4 up 1 down) (Figure 9.12d). Since the graph is meant for the extra warp, the ground and figure parts of the edited graph are given selective contouring (top and bottom) for 2 pixels with two different colours (Figure 9.12e). Then, the weaves are mapped with the respective colours (Figure 9.12f). Finally, all the figure colours are converted into one colour (red). The ground contour colour is converted to ground colour and the figure contour colour is converted to figure colour. With this, the final jacquard graph for the figured extra warp is ready. This graph is without any long vertical end floats. It is without any binding marks on the top and bottom edges. Hence, the boundaries of the ground and figure are retained without any distortion as shown in Figure 9.12g.

9.6.2　Extra Weft-Figured Graph

To prepare the figured extra weft graph, the contouring is applied in the left and right directions instead of the top and bottom directions as shown in the figured extra warp graph preparation. Hence, the procedure to prepare the figured extra weft graph is the same as the procedure followed to prepare the figured extra warp graph as illustrated in the flowchart given in Figure 9.11, with only one change in step 4. That is, in step 4, instead of applying top and bottom contouring, left and right contouring has to be applied. Figure 9.13a shows the left and right contouring for two pixels in the ground and figure. Figure 9.13b shows the mapping of binding weave marks both on the ground and figure. Finally, all the figure colours are converted into one colour (red). The ground contour colour is converted to ground colour and the figure contour colour is converted to figure colour. With this, the final jacquard graph for the figured extra weft is ready. This graph is without any long horizontal pick floats. It is without any binding marks at the left and right edges. Hence, the boundaries of ground and figure are retained without any distortion as shown in Figure 9.13c.

9.6.3　Single Cloth Figured Graph

The flowchart illustrated in Figure 9.14 gives the steps to prepare the figured single cloth graph using top-bottom and left-right contour options. First, the top-bottom contoured graph is opened (Figure 9.12e). Then, the left-right contoured graph (Figure 9.13a) is opened as a layer over the top-bottom contoured graph. This is shown in Figure 9.15a. Now, the contour colours magenta and cyan are made transparent. When two files are merged, the top-bottom contouring and left-right contouring get combined; only the two corner pixels in the boundary of the ground and figure are left in magenta and cyan colours, respectively, as shown in Figure 9.15b. Then, the ground and figure weaves are mapped with the ground and figure colours, respectively, as shown in Figure 9.15c. Finally, all the figure colours are converted into one colour (red), and all the binding weave mark colours in the figure parts are converted into one colour (green). The ground contour colour is converted to ground

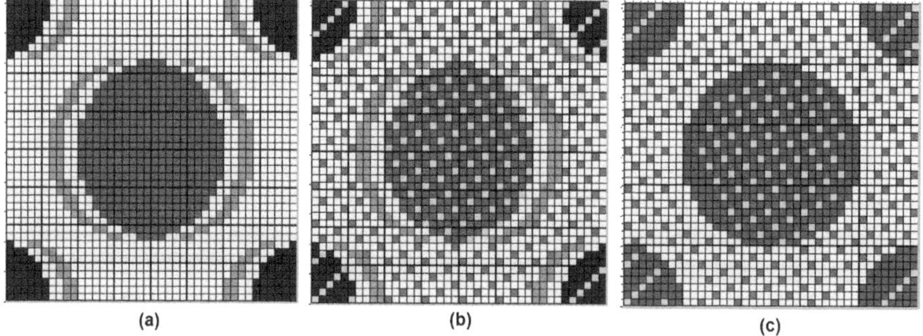

FIGURE 9.13　Preparing the figured extra weft graph using the selective contour option (left and right): (a) left and right contouring in the ground and figure, (b) mapping ground and figure weaves, and (c) final float-controlled figured extra-weft graph.

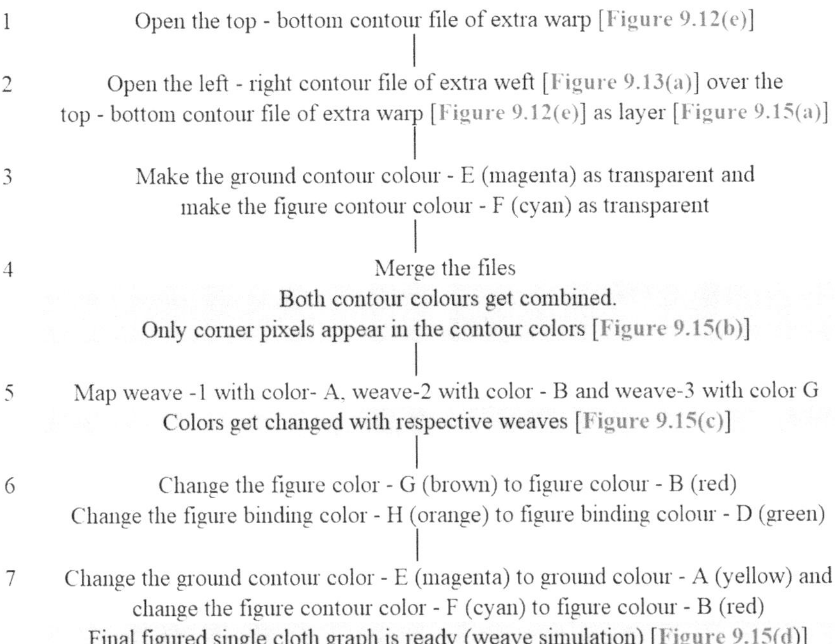

FIGURE 9.14   Flowchart showing the algorithm to prepare figured single cloth graph using the selective top—bottom and left—right contour options.

colour, and the figure contour colour is converted to figure colour. With this, the final jacquard graph for the figured single cloth is ready. This graph is without any long vertical end floats and horizontal pick floats. It is without any binding marks in the stepping corners. Hence, the boundaries of the ground and figure are retained without any distortion as shown in Figure 9.15d.

9.7   PREPARING THE EXTRA WEFT-FIGURED GRAPH WITH PLAIN PICKS

In extra weft weaving, one plain pick is woven for each extra weft figure pick. When mechanical jacquard is used for weaving, the ends are operated in plain order for weaving the plain pick, either using 2 healds or plain cards. When plain cards are used, they are punched without the aid of any graph. Extra weft figuring cards are punched from the extra weft-figured graph. Then, these two sets of cards are laced in a 1:1 order and taken for weaving. When an electronic jacquard is used without healds for operating the ends in plain order, the extra weft-figured graph is prepared with plain picks and extra weft picks in 1:1 order. The following procedure explains how to prepare the extra weft-figured graph with plain picks from the final float-checked extra weft-figured graph prepared, as shown in Figure 9.13c.

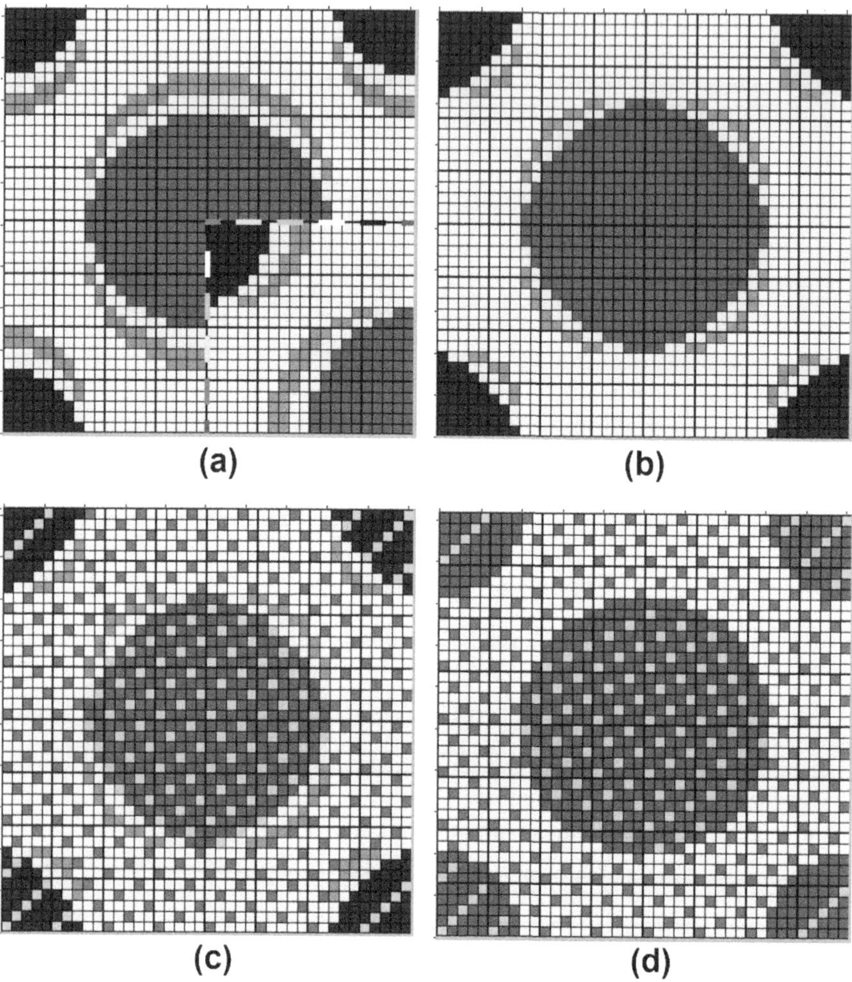

(a)   **(b)**

(c)   **(d)**

FIGURE 9.15 Preparing the figured single cloth graph using selective top-bottom and left-right contour options: (a) opening left—right contoured file as a layer over top—bottom contour file, (b) contouring only on corner pixels is visible after merging the file, (c) mapping weaves on the ground and figure, and (d) final float-controlled figured single cloth graph.

i. Open the final float-checked extra weft-figured graph on 40×40 (Figure 9.13c). Convert all the warp colours (red and blue) into one colour (red). Convert all the weft colours (yellow and green) into one colour (white). The graph thus converted is shown in Figure 9.16a.

ii. Prepare a graph of 40×40 size with the plain weave and save it as a plain weave graph.

iii. The extra weft-figured graph is on 40 ends $\times$ 40 picks, and the plain weave graph is on 40 ends $\times$ 40 picks. Therefore, the extra weft graph combined

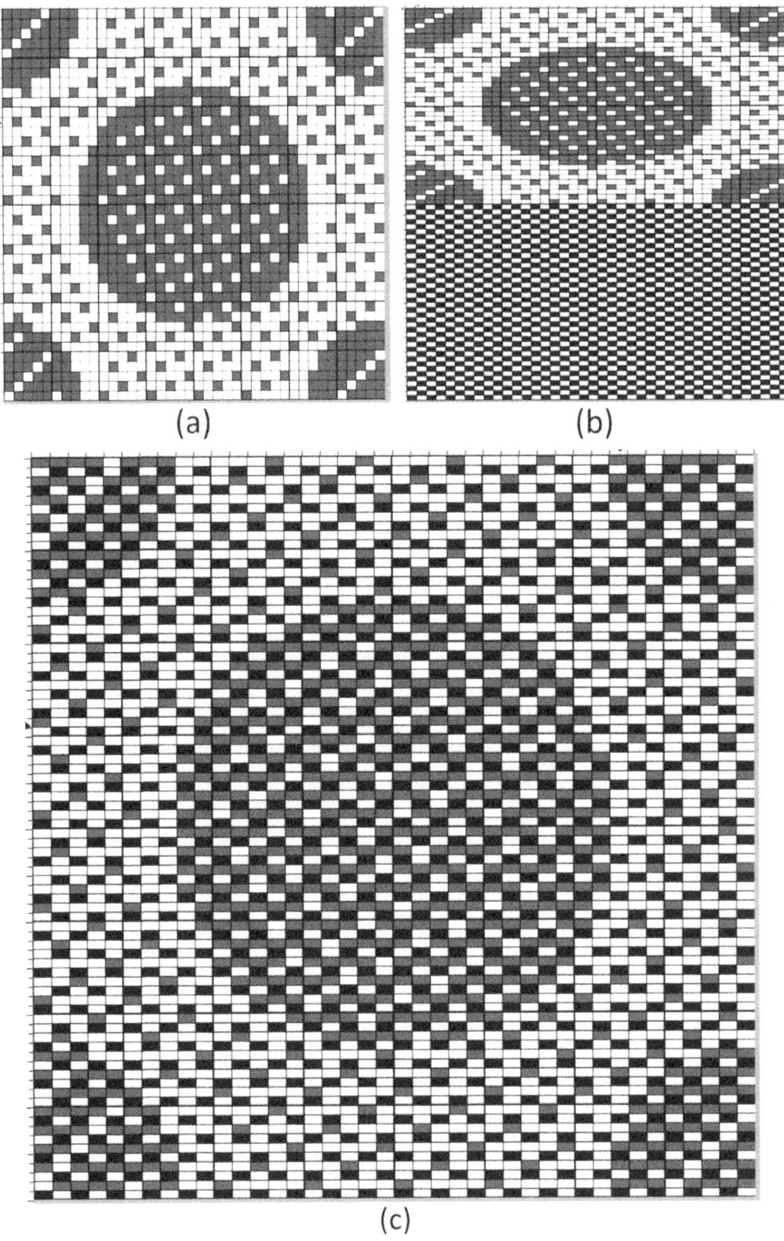

FIGURE 9.16 Preparing the extra weft-figured graph with plain picks: (a) the final float-controlled figured extra weft graph (40 × 40) in two colours, (b) opening the sectional graph (40 × 80); opening the plain graph (40 × 40) at the bottom section of the sectional graph; and opening the extra weft-figured graph (40 × 40) at the top section of the sectional graph, and (c) the final extra weft-figured graph on 40 × 80 with plain and extra weft picks in 1:1 order obtained after merging.

with plain picks in 1:1 order will be 40 ends × 80 picks (40 plain picks + 40 extra-weft picks). Hence, open a new graph of 40 ends × 80 picks.

iv. When the extra-weft picks combine with plain picks, the picks per inch become double the ends per inch. Hence, set the picks per inch to double the ends per inch. This will make the 40 × 80 new graph to form a square, as seen in Figure 9.16b. Also, make the graph count as 5 × 10, to match the ratio of 40 × 80. This will result in the thick grid lines forming a square. This is also seen in the graph given in Figure 9.16b.

iv. Using the 'Divide horizontal section' option, divide the graph into two horizontal sections, each of 40 × 40, one at the bottom and the other on the top.

v. Open the plain weave graph (40 × 40) as a layer over the bottom section of the 40 × 80 graph and open the extra-weft figure graph (40 × 40) as a layer over the top section of the 40 × 80 graph, as shown in Figure 9.16b.

vi. Using the 'Merge horizontal section' option, merge both sections into one section. By doing so, the 40 plain picks in the bottom section and the 40 extra-weft figure picks in the top section automatically get merged in 1:1 order. Figure 9.16c shows the final extra-weft-figured graph of 40 × 80 obtained after merging with plain and extra-weft figure picks in 1:1 order.

In the examples discussed in this chapter, particular colours are used for easy cross-reference and understanding. The ground and figure colours are taken as one/ two in the examples. The number of colours can be taken as equal to the number of weaves. The colours used in the edited graph are other than black and white. While using the all-directional contour option, the single connecting pixel is taken (not the single nose-touching pixel). It gives contouring for one pixel in the top, bottom, left, and right border. It also gives contouring for two pixels in the corners. While using the selective directional contour option, two nose-touching pixels are taken (not the two connecting pixels). It gives contouring for two pixels in all directions. Two different contour colours are used: one for all the ground colours, and the other for all the figure colours. The icons and names of the tools and options shown in the examples are general. It is not about any software. These may differ from software to software. When the designers use particular software, tools and options may be selected and used as per the purpose and functions performed as described in the examples. The number of pixels taken in the 'measure' option to identify the floats depends upon the float in the weave used. Commonly two or three pixels more than the float length in the binding weave are taken. 9 or 10 pixels are taken while using a threads binding weave (7 floats). 6 or 7 pixels are taken while using a five-thread binding weave (4 floats). In preparing the single-cloth graph using all-directional contour options, the warp floats are identified first and then the weft floats. It can be vice versa also. Similarly, while preparing the single-cloth graph using the selective directional contour option, the top-bottom contour graph is opened first over the left-right contour graph. It can be vice versa also. In all the algorithms, the action performed is shown in white background and the result of the action is highlighted with yellow colour. The procedure derived for the figured extra warp graph can be used for the figured warp-backed graph. The procedure derived for the figured extra-weft graph can be used for the figured weft backed graph. Similarly, the extra-weft-figured graph can be obtained from the extra warp graph by rotating the final extra warp-figured graph by 90°. Likewise, the extra-weft-figured graph prepared with plain picks can be used as an extra warp-figured graph with plain ends by rotating the final extra-weft-figured graph, by 90°.

BIBLIOGRAPHY

1. Grosicki, Z. J. (2004). Construction and development of jacquard design. In *Watson's Textile Design and Colour* (pp. 208–230). Woodhead Publishing Limited, Cambridge.
2. Grosicki, Z. J. (2004). Figuring with extra threads. In *Watson's Advanced Textile Design* (pp. 11–41). Woodhead Publishing Limited, Cambridge.
3. Panneerselvam, R. G. and Prakash, C. (2022). Study on the float-control weaves application algorithms of computer-aided jacquard graph designing for different figured fabrics. *Journal of the Textile Institute 2023*. 114(11), 1727–1739. DOI: 10.1080/00405000.2021.2023956.

10 Figured Graph Simulations, Draping, Graph Printing, Punching, and Lacing

10.1 FIGURED GRAPH SIMULATIONS AND DIGITAL OUTPUT

During the process of developing the jacquard graph from the motif, the designers must save and take (i) the printouts of the different motif stages, which are called simulations, (ii) the digital output of the graph with marks and blanks, and (iii) the graph paper printout of the digital graph.

The simulation option is a very useful feature available in CAFGD software. By using the simulation option, it is possible to visualize how the edited motif looks on the screen of the computer and will appear as fabric with proper repeat size after weaving. In figured graph making, there are three different stages for simulating the design and graph.

i. After editing the graph with a proper outline, the figure and ground parts are filled with the required colours and then visualized. This stage is called the edited 'colour graph/motif simulation'. This simulation on the screen is called a real-scale colour graph/motif simulation if the repeat of the graph simulation on the screen is equal to the repeat size when it is woven as fabric. If this real-scale colour graph/motif simulation on the screen is printed on paper in the same size, it is called real-scale colour graph/motif simulation printing. This motif colour simulation stage gives an idea of only the figure and ground areas. It does not give any knowledge about the weaves applied to the figure and ground and the yarn used.

ii. In the edited motif, the figure and ground parts are filled with the number of colours as per the number of weaves required to apply (one colour for one weave), and the weaves are mapped for all the colours. After weave mapping, the design will appear with actual weaves in different figure and ground parts. The required warp colour is given for the warp up (marks) and the weft colour for the weft up (blanks), and then, the graph design is visualized. This is called the 'figured weave graph simulation' stage. This simulation on the screen is called a real-scale figured graph weave simulation if the repeat of the simulation on the screen is equal to the repeat size when it is woven as fabric. If this real-scale figured weave graph simulation on the screen is printed on paper in the same size, it is called real-scale figured

DOI: 10.1201/9781003441205-10

weave graph simulation printing. This simulation gives an idea of the actual weave effect with warp and weft colours. However, it does not give any idea about the type of warp and weft yarns and their interlacement.

iii. For the figure and ground parts of the figured weave graph, the interlacement of warp and weft yarns, as appears on the face of the fabric, is assigned. Now, the design appears with the required weave and interlacement in a three-dimensional view, as seen in the real fabric. This stage is called the 'figured fabric simulation'. This simulation on the screen is called a real-scale figured fabric simulation if the repeat of the simulation on the screen is equal to the repeat size when it is woven as fabric. If this real-scale figured fabric simulation on the screen is printed on paper in the same size, it is called real-scale figured fabric simulation printing.

10.1.1 Figured Weave Graph Simulation

For example, consider that the designer is developing a jacquard graph for weaving the interchanging plain double cloth (IPDC) in four colour effects. The repeat size of the motif on paper is $2'' \times 2''$. Figure 10.1a shows the 2×2 repeats of the motif on $4'' \times 4''$. The ends and picks per inch are 48. The jacquard capacity used is 480. The warp colouring order is 1 light colour end: 1 dark colour end (LE: DE). The weft colouring order is also 1 light colour pick: 1 dark colour pick (LP: DP).

The total ends and picks of the IPDC-figured graph are taken as 96×96 ($2'' \times 48$; $2'' \times 48$). The motif is scanned to have the repeat in half of the total ends and picks of the graph size, that is 48 ends $\times$ 48 picks. The scanned graph is edited as per the motif. After editing, the graph size is made into 96×96 by doubling each end and

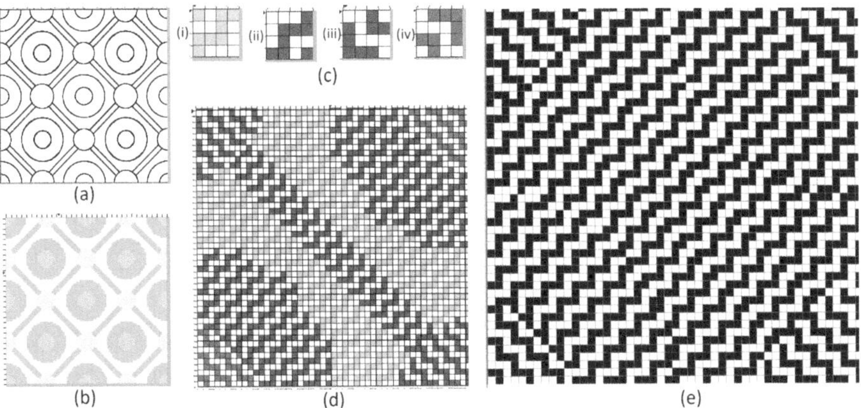

FIGURE 10.1 Preparing digital figured weave graph: (a) four repeats of figured motif, (b) four repeats of edited figured graph simulation of the motif with different colours on figure and ground parts, (c) four weaves (in four colours) for four colour effects of interchanging plain double cloth (IPDC), (d) figured weave graph mapped with four weaves shown in (c), and (e) figured weave graph with four weaves converted into black and white for electronic jacquard weaving and electronic card punching.

pick by using the multiply option. This method enables the designer to apply the weave marks of IPDC perfectly, which are based on the 1:1 colouring order for both warp and weft. Next, the figure and ground parts are given four different colours to apply four different weave marks to get four different colour effects. The ground is given yellow colour to apply the weave of the first colour effect. The outer part of the centre circle and the diamond lines are given orange colour to apply the weave of the second colour effect. The circle at the diamond line joining point is given light blue colour to apply the weave of the third colour effect. The inner circle of the centre circle is given light purple colour to apply the weave of the fourth colour effect. After applying the colours, the correct repeating of the design and colours is verified by having 2 × 2 repeats on 192 × 192. On the screen, when this graph design (pixels) is zoomed out without the graph to 4″ × 4″ size (2 repeats × 2″ per repeat), it is the real-scale edited colour graph/motif simulation. When the designer takes the printout of this graph design on 4″ × 4″ size on paper as shown in Figure 10.1b, it is the colour simulation printout of the figured graph developed as per the motif. This colour simulation just represents how the figure and ground parts look in four colour effects in the cloth.

Then, the designer has to derive the four weaves of the IPDC to get the four colour effects described below:

i. First colour effect: Light colour ends interlace with light colour picks in plain weave on the face, and dark colour ends interlace with dark colour picks in plain weave on the back. It is represented as [(LE + LP) / (DE + DP)].

ii. Second colour effect: Dark colour ends interlace with dark colour picks in plain weave on the face, and light colour ends interlace with light colour picks in plain weave on the back. It is represented as [(DE + DP) / (LE + LP)].

iii. Third colour effect: Light colour ends interlace with dark colour picks in plain weave on the face, and dark colour ends interlace with light colour picks in plain weave on the back. It is represented as [(LE + DP) / (DE + LP)].

iv. Fourth colour effect: Dark colour ends interlace with light colour picks in plain weave on the face, and light colour ends interlace with dark colour picks in plain weave on the back. It is represented as [(DE + LP) / (LE + DP)]. The weaves of these four colour effects are shown in Figure 10.1c. The marks (end up) of the first weave (i) are in green, the second weave (ii) are in red, the third weave (iii) are in blue, and the fourth weave (iv) are in purple. These weaves are saved under different file names.

Now, the designer has to apply these four weaves on four different colours on the 96 × 96 edited graph (one repeat). The four weave mark files are taken as a brush and applied over the colours in the edited graph using the bucket fill tool. The first weave file (marks in green colour) is applied over the yellow colour area in the edited graph to get the first colour effect [(LE + LP) / (DE + DP)]. The second weave file (marks in red colour) is applied over the orange colour area in the edited graph to get the second colour effect [(DE + DP) / (LE + LP)]. The third weave file (marks in blue colour) is applied over the light blue colour area in the edited graph to get the third colour effect [(LE + DP) / (DE + LP)]. The fourth weave file (marks in purple colour) is applied over the light purple colour area in the edited graph to get the fourth colour effect [(DE + LP) / (LE + DP)]. After completing the application of weave marks, the four different figure parts of the graph will appear with four different colour weave marks on white ground, which is the 'figured weave graph in colour'. The 48 × 48 graph shown in Figure 10.1d is the bottom left part of the 96 × 96 'figured weave graph in colour' thus obtained. The 96 × 96 graph prepared is for the jacquard of 480 hooks. Hence, the graph is made to repeat five times in the end-way to have the graph on 480 ends × 96 picks. Then, all the colours of the weave marks (ends up) are changed to black marks, as shown in Figure 10.1e, which is the 'figured weave graph in black and white'. This figured graph in black-and-white pixels is for the electronic jacquard weaving or electronic card punching machine.

In the jacquard, apart from the main hooks used for controlling the figuring ends, there are idle hooks, extra hooks that are used to control the selvedge ends, healds, and mechanisms like colour selection and take-up control. Hence, after completing the main graph as above, the graphing for these extra hooks is also added, corresponding to the ends, healds, and mechanisms that these hooks control. After completing the graphing for the hooks, the digital graph is saved in the suitable format suggested for operating the electronic jacquard or electronic punching machine.

The designers must keep in mind that the electronic jacquard produces the top-closed shed. The digital black mark on the graph lowers the hook, and the blank on the graph keeps the hook on top. Hence, the actual weave effect marked on the graph is woven at the backside while weaving. If it is required to have the actual weave effect marked on the graph, to weave on the face side, the figured graph is made negative. That is, black marks are made as white and white marks are made as black. The graph for the other hooks is also made accordingly.

10.1.2 Figured Fabric Simulation

The designer can also do a simulation of the jacquard graph with the actual colouring order of the ends and picks used while weaving the fabric. Consider that the colouring order of the ends is one yellow (light) and one black (dark). It can be marked on an 8 × 8, 4 × 4, or 2 × 2 graph and saved as a 'warp colouring order' file. The colouring order of the picks is one blue (light) and one maroon (dark). It can be marked on an 8 × 8, 4 × 4, or 2 × 2 graph and saved as a 'weft colouring order' file. The colouring order of ends marked on an 8 × 8 graph (i) and the weft colouring order marked on 8 × 8 (ii) are shown in Figure 10.2a. The colouring order file is opened as a brush and applied over the figured graph in black and white (Figure 10.1e). The warp colouring order is applied for the black colour (end up), and the weft colouring order is applied for the white colour (end down). The graph now obtained is the 'figured weave graph with yarn colour'. The 48 × 48 graph shown in Figure 10.2b is the 'figured weave graph in yarn colour'. It is the bottom left part of the 96 × 96 graph.

Figure 10.2c shows the four weave marks of the four colour effects on 4 × 4 with ends and picks marked as per the 1:1 colouring order of ends and picks. These are the four weave marks of the figured graph shown in Figure 10.2b. In each of these weave marks, it is observed that, apart from the interlacing of face ends and face picks in plain weave, the colour dots of the back ends and back picks are also visible. In the

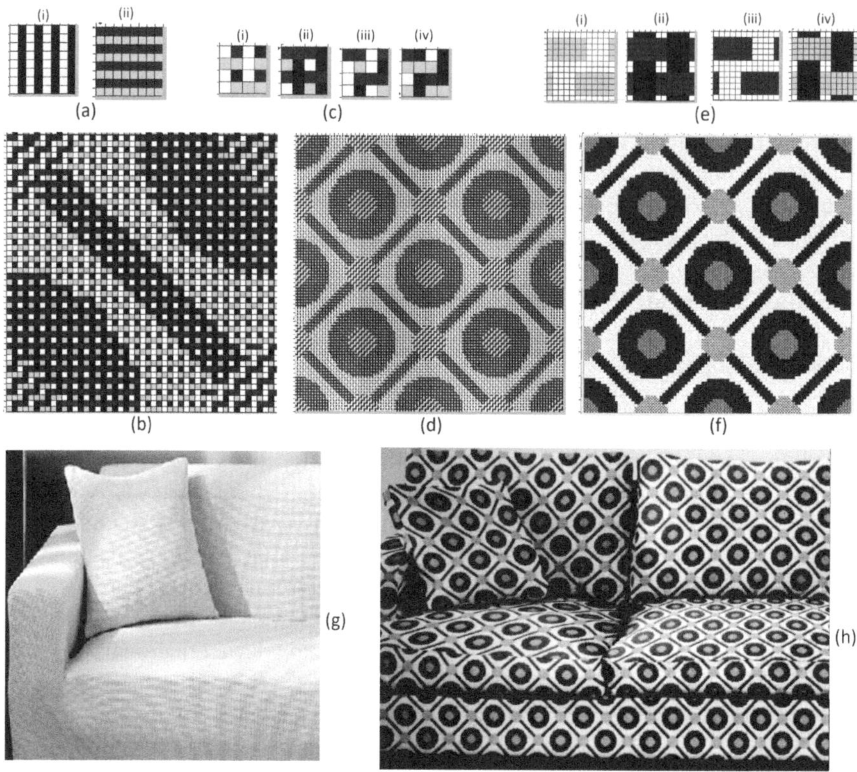

FIGURE 10.2 Preparing fabric simulation and fabric draping simulation: (a) 1:1 warp and weft colouring order, (b) figured weave graph mapped with warp and weft colours as shown in (a), (c) four weaves of IPDC with warp and weft colours as shown in (a), (d) simulation of figured weave graph shown in (b), (e) face end and face pick interlacement of four weaves of IPDC, (f) fabric simulation mapped with the interlacements shown in (e), (g) a sofa with plain cloth taken for draping the fabric simulation, and (h) sofa after draping the fabric simulation shown in (f).

first weave (i), the yellow face ends (1, 3) and light blue face picks (1, 3) are interlaced in plain order on the face side. Apart from that, the black back ends (2, 4) and maroon back picks (2, 4) are visible as dots, which are the interlacements of these ends and picks on the backside, as shown on 4×4 in Figure 10.2c (i). Similar interlacing and the dots seen in the other three weaves are also shown in Figure 10.2c (ii–iv).

The repeat of this figure weave graph with yarn colour is made into 2×2 repeats of 192×192 size. On the screen, when this graph design (pixels) is zoomed out without graph to $4'' \times 4''$ size (2 repeats $\times$ 2" per repeat), it is the 'fabric simulation' of the figured graph of the given motif. When this simulation is printed on paper to $4'' \times 4''$ size, it is the fabric simulation of the figured graph on the paper, as shown in Figure 10.2d. This simulation is the prelude to the final fabric, representing how the figure and ground parts more or less look in four colour effects when it is woven as IPDC.

However, in the actual IPDC, the plain fabrics formed by the interlacing of face ends and face picks are only visible on the face side. The face plain fabric hides the interlacing of the back ends and the back picks, which form the plain fabric at the backside. Hence, to create the real fabric simulation of the IPDC, it is essential to create the interlacement of only face ends and face picks in all the weaves of the IPDC on the required number of ends and picks per repeat. For example, the repeat is taken on 12 × 12, with the thickness of the ends and picks equal to four and the gap between the ends and picks equal to two. The four plain interlacements thus created on the face side—(LE + LP), (DE + DP), (LE + DP), and (DE + LP)—are shown serially in Figure 10.2e (i–iv). The edited motif on 48 × 48 size (Figure 10.1b) is made to 576 × 576 by multiplying each pixel by 12 × 12. This is because the plain interlacement of each weave is created on 12 × 12. Then, each interlacement is opened as a brush and applied over the corresponding colours of the 576 × 576 graph. The first interlacement (LE + LP) is applied over yellow. The second interlacement (DE + DP) is applied over the orange. The third interlacement (LE + DP) is applied over light blue. The fourth interlacement (DE + LP) is applied over purple. The graph now obtained is the final 'figured fabric simulation'. The repeat of this graph is made into 2 × 2 repeats of 1152 × 1152 size. On the screen, when this graph design (pixels) is zoomed out without graph to 4″ × 4″ size (2 repeats × 2″ per repeat), it is the 'fabric simulation' of the figured graph. When this graph design is printed on paper to 4″ × 4″ size, it is the fabric simulation of the figured graph on the paper, as shown in Figure 10.2f. This is the final fabric simulation, representing how the figure and ground parts look exactly in four colour effects when it is woven as IPDC.

10.2 FIGURED FABRIC DRAPING SIMULATION

Most of the woven figured cloths are either used as stitched dress materials for humans or used as furnishings for furniture. The colour pattern and weave structure used in the cloth must suit the end use. Therefore, the figured graph designer must have a perception about how the colour pattern and the figured weave structure developed look when it is used as a dress or furnishing. The CAFGD software helps the designer to carry out this work more easily and effectively. The software has folders with photographs of the drawing room, bedroom, and car seat, as well as photographs of women, men, and children with different styles of dressing. The required photograph file is opened on one side of the window. The fabric simulation file is opened on the other side. Over a particular dress or piece of furniture, the selected design file is made to combine or merge by the process of 'dragging and dropping'. By doing so, the dress or the furniture gets changed as per the new fabric simulation mapped. This process of dragging and dropping the fabric simulation over the selected object to see its appearance is called 'texture mapping' or 'fabric simulation draping'. By doing so, the figured graph designer gets an idea of the final appearance of the design in its end use. For example, the figured fabric simulation of 4″ × 4″ size developed by the designer on the screen shown in Figure 10.2f is opened on one side of the window, and the sofa with a plain cloth cover shown in Figure 10.2g is opened on the other side of the window. The figured fabric simulation is dragged and dropped over the sofa. Figure 10.2h shows the sofa with the figured fabric after dragging and dropping. It is called the 'draping simulation' of the fabric simulation. Now, the designer can visualize the appearance of the figure and ground parts of the motif with double-cloth four colour effects and how it suits when furnished over the sofa. The designer can change the colour combination of warp and weft used in the fabric simulation and drag and drop the new file to ascertain the suitability of the subsequent colour combinations compared with the first one. All the simulations created on the screen and printed on the paper as described above help the designer to visualize the figured motif up to its end use only by completing the fabric simulation on the screen, before going for actual weaving. This draping helps the designer to be satisfied with the simulation created and also send it for approval.

10.3 FIGURED GRAPH PRINTING FOR MANUAL PUNCHING

After doing all the simulations, the full-figured colour weave graph on 480 × 96 (Figure 10.1d shows a part of the full graph on 48 × 48) is printed as a figured graph on paper for manual punching. For this, the graph count and the shape of the weave marks on which each colour pixel has to be printed are decided. The following are the important parameters to be checked while taking graph printouts on paper:

   i. Size of graph and graph count: The total number of ends and picks can be printed on the required size and the graph count. For example, a 480 × 96 graph can be printed on a 48″ × 9.6″ size. The resulting graph printout will have 10 vertical and 10 horizontal spaces in each square inch space (10 × 10 graph count—because 480/48″ = 10; 96/9.6″ = 10). When the same graph is printed on 40″ × 8″, the resulting graph printout will have 12 vertical and 12 horizontal spaces in each square inch space (12 × 12 graph count—because 480/40″ = 12; 96/8″ = 12).
   ii. Thickness of the graph lines: There is an option to select the required graph count. According to the selected graph count, the thick lines are printed. There is also an option to increase or decrease the density of the thick and thin lines of the graph to be printed as per the requirement.
   iii. Colour and size of the weave marks: It is possible to change the colour of the weave marks on the ground and figure areas as per the punching requirement. It is also possible to increase or decrease the amount of colour filling on each square block of the graph. By giving 70% filling, the colour is filled 70% in the middle of each square block, and 30% is left as white blank surrounding the colour. By giving 40% filling, the colour is filled only for 40%, and 60% is left blank. 40% of filling consumes less colour ink than 70% of filling, and the graph can be printed at a cheaper cost.
   iv. Different symbols for different colours: When the graph printout is taken on the black-and-white printer, it is not possible to get different colours on the figure and ground weaves as seen on the monitor. To overcome this difficulty, an option is provided in the printing parameters. By using this option, different symbols like 0, X, *, <, >, /, and \ can be assigned to different colours of the weaves, and the graph is printed with these symbols in different figure and ground parts.

    v. Numbering of hooks and picks: It is possible to assign and print the numbering of hooks and picks, starting from the required side of the graph. The numbering can be started from either bottom left or bottom right or top left or top right as per the harness building. The hook and pick numbering is useful for counting and reading the total number of hooks and picks on the printed graph from the required side while punching.

   vi. Margins: As per the paper size used, margins of the required size can be left on all four sides of the paper. When the graph is printed using an A3 or A4 size printer, these margins help to join the papers one after the other to get the full graph sheet. Apart from these options, it is also possible to print page numbers and graph particulars.

Consider that the graph is printed for manual card punching, and the cards are used for the 480 hooks (240 hooks × 2) with 8 hooks per short row. Hence, the graph count is selected as 8 × 8 (ends per inch is equal to picks per inch). The shape of the marks for each colour in the weave graph shown in Figure 10.1c is decided. For example, 'square' marks for green pixels, 'circle' marks for red pixels, 'x' marks for blue pixels, and 'slash' marks for purple pixels are decided, as shown serially (i–iv) in Figure 10.3a. For printing the graph, the CAFGD software has the option to assign the graph count and shape of the weave mark for each colour decided as above. With these inputs, when the print graph option is clicked, the print graph appears on the screen with graph lines on 8 × 8 count and the weave marks in different shapes in black. The graph size is 480 ends × 96 picks. Hence, to print the graph with ten ends and ten picks per inch, the print size is given as 48″ × 9.6″ (480/10, 96/10). As per the printer size used for printing, the graph gets printed on many pages or one page. A part of the punching graph on 48 × 32 (from the graph shown in Figure 10.1d),

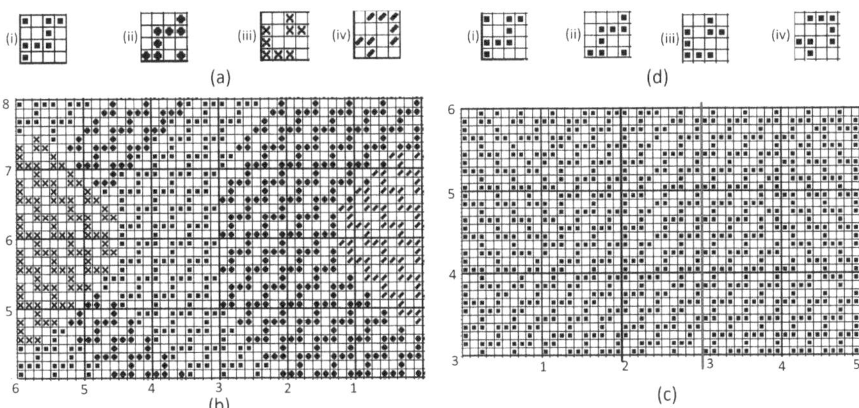

FIGURE 10.3 Figured weave graph printing for card punching: (a) four weaves of IPDC with four different styles of marks, (b) figured weave graph of 48 × 32 size printed as graph on paper with an 8 × 8 graph count, with four weaves shown at (a), (c) figured weave graph of 50 × 30 size printed as graph on paper with a 10 × 10 graph count, with four weaves shown at (d), and (d) four weaves of IPDC with one style of marks.

thus printed, is shown in Figure 10.3b, with the graph count of 8 × 8 and the marks of the different weaves in different shapes. The sequence number of the thick lines of the graph count also gets printed at the margin of each page. 1, 2, 3,... numbers printed near each vertical thick line in the end-way represent the counting of 8, 16, 24,... ends. Similarly, 1, 2, 3,... numbers printed near each horizontal thick line in the pick-way represent the counting of 8, 16, 24,... picks. The printed graph shown in Figure 10.3b with the vertical thick lines numbering from 1 to 6 indicates that the total ends printed are 48 (from 1 to 48). The horizontal, thick lines numbering from 5 to 8 indicate that the total picks printed are 32 (from 33 to 64). When the A4 size printer is used, the total graph gets printed on seven pages. After printing, the pages are pasted to have the graph paper size of 48″ × 9.6″. The vertical and horizontal thick line numbering printed at the margins helps to paste the papers in proper sequence. A part of another punching graph (50 × 30) is shown in Figure 10.3c. This punching graph is printed with a 10 × 10 graph count and all four different weave marks in one shape (square), as shown serially (i–iv) in Figure 10.3d. This printing with a 10 × 10 graph count is useful for long-row punching (30 hooks per long row × 8 long rows = 240 hooks) by drawing the red colour vertical line over the thick black vertical line after each set of 30 ends, as shown in the graph (Figure 10.3c).

Table 10.1 shows the consolidated list of different forms of figured graphs, their simulation, and printouts with reference figure numbers and various representations of four different weaves of figured IPDC with four colour effects.

10.4 CARD PUNCHING AND LACING

The figured weave graph printed on paper is now used for punching the cards. Different punching machines used for card punching are (i) manual and mechanical card punching, which includes hand punching plates and pedal-operated mechanical piano card punching machines, and (ii) electronic automatic punching machines.

10.4.1 Manual and Mechanical Card Punching

i. Hand punching plate: Punching the cards manually using hand punching plates is the oldest method of card punching being used for the small-capacity jacquards (up to 120 hooks) used in the decentralized handloom sector. Even today, most of the handloom sectors engaged in saree weaving with jacquard use hand punching methods. The reason is that in most of the saree weaving with jacquard, long-row harness building is in practice. Therefore, punching is also done long row wise using hand punching plates. Mechanically operated piano card punching machines punch the cards only in short rows. That is why piano card-cutting machines cannot replace hand punching plates.

Another question to be asked is why only long-row-wise harness building is mostly preferred in jacquard saree weaving looms; why not use a short-row-wise harness building so that it is possible to use a mechanical punching machine? The reason is that in saree weaving with an extra weft

TABLE 10.1

Consolidated List of Different Forms of Figured Graphs, Their Simulation, and Printouts with Reference to Figure Numbers and Various Representations of Four Different Weaves of Figured IPDC with Four Colour Effects

Fabric variety—figured 'interchanging plain double cloth' (IPDC) with four colour effects.

Ends and picks per inch—48.   Motif size—$2'' \times 2''$.

Jacquard capacity used—480.

Warp colour—yellow (light) and black (dark) in 1:1   order (Figure 10.2a).

Weft colour—blue (light) and maroon (dark) in 1:1   order (Figure 10.2a).

Different forms of figured graphs, their simulation, and printouts	Reference figure no.	Four colour effects			
		[(LE + LP) / (DE + DP)]	[(DE + DP) / (LE + LP)]	[(LE + DP) / (DE + LP)]	[(DE + LP) / (LE + DP)]
Figured motif on paper	Figure 10.1a	-	-	-	-
Edited colour graph simulation	Figure 10.1b	Yellow	Orange	Blue	Purple
Four weaves	Figure 10.1c	Ends up in green	Ends up in red	Ends up in blue	Ends up in purple
Figured weave graph in colour	Figure 10.1d				
Figured weave graph in black and white (digital output for electronic weaving and electronic punching)	Figure 10.1e	In all the weaves, ends up is in black			
Figured weave graph in warp and weft colouring order	Figure 10.2b	All the weaves are with 1:1 colouring order of ends and picks			
Four weaves	Figure 10.2c				
Figured weave graph simulation	Figure 10.2d				

(Continued)

TABLE 10.1 (*Continued*)

Consolidated List of Different Forms of Figured Graphs, Their Simulation, and Printouts with Reference to Figure Numbers and Various Representations of Four Different Weaves of Figured IPDC with Four Colour Effects

Fabric variety—figured 'interchanging plain double cloth' (IPDC) with four colour effects.

Ends and picks per inch—48.　Motif size—2″ × 2″.

Jacquard capacity used—480.

Warp colour—yellow (light) and black (dark) in 1:1　order (Figure 10.2a).

Weft colour—blue (light) and maroon (dark) in 1:1　order (Figure 10.2a).

Different forms of figured graphs, their simulation, and printouts	Reference figure no.	Four colour effects			
		[(LE + LP) / (DE + DP)]	[(DE + DP) / (LE + LP)]	[(LE + DP) / (DE + LP)]	[(DE + LP) / (LE + DP)]
Four interlacements	Figure 10.2e	Interlacement of only LE + LP (yellow + blue)	Interlacement of only DE + DP (black + marron)	Interlacement of only LE + DP (yellow + marron)	Interlacement of only DE + LP (black + blue)
Figured fabric simulation	Figure 10.2f				
End-use product	Figure 10.2g	-	-	-	-
Draping simulation with fabric simulation	Figure 10.2h	-	-	-	-
Four weaves	Figure 10.3a	Ends up in square mark	Ends up in circle mark	Ends up in 'x' mark	Ends up in '/' mark
Figured graph printed as graph on paper with 8 × 8　graph count (for manual punching)	Figure 10.3b				
Four weaves	Figure 10.3d	All the weaves are with the 'square marks' for the ends up			
Figured graph printed as graph on paper with 10 × 10　graph count (for manual punching)	Figure 10.3c				

weaving style, the design area is greater than the ground area. Hence, the number of continuous holes punched is also higher. When continuous holes are punched in the short-row-wise method, the card becomes weaker; hence, there are possibilities of frequent card breaking while weaving. When continuous holes are punched in a long row, the card will not become that weak, and hence, the possibility of the card breaking is also less. For these reasons, the handloom jacquard saree weaving centres use only hand punching plates for card punching.

In hand punching, first, the expert designer keeps the card in between the two perforated plates, reads the graph, and puts ink marks in the required places where holes are to be made. Then, the punching person keeps the marked card again in between two perforated punching plates and punches the marked places by using a single punch with a hammer. A separate punch is used for making peg holes. The rate of punching by using a hand punching plate is 20–30 cards per hour (120 hooks) because this process is more laborious.

ii. Mechanical piano card punching machines: In the handloom and power loom sectors, jacquards constructed with short-row-wise harnesses are used to weave figured fabrics. Mechanical piano card-cutting machines move and punch the cards short row wise. Therefore, to punch cards for these jacquards, mechanically operated piano card punching machines are used. In this machine, the card is held in a carriage, and the punching person operates the pedals with his legs. Each pedal operation drags the card backwards by one short row and moves the headstock up and down with punches to punch the card. The punching person reads the marks in the punching graph and operates the keys in the headstock with his fingers as per the marks. The keys, in turn, regulate the punches to punch the card. Each card is fed manually, and after punching, it is removed manually. The rate of punching of the piano card-cutting machine is 25–30 cards per hour (240 hooks).

10.4.2 Electronic Automatic Card Punching

Nowadays, after the introduction of CAFGD software, electronic automatic card punching machines have become popular. Figure 10.4a shows the bird's eye view, Figure 10.4b shows the front view, and Figure 10.4c shows the side view of the electronic automatic card punching machine. The punching machine is attached to the monitor and the processor. The digital graph file prepared from the CAFGD is saved in the processor. The punching machine works as per the digital graph saved in the processor. The punching machine performs the following work automatically: take the card for punching and feed it in between two rollers. These rollers move the card short row wise as per the pitch. The headstock moves up and down with punches controlled by the keys. The processor electronically controls the keys as per the marks of the graph design. The monitor shows the details of three cards (Figure 10.4d). The middle one is the card that is under punching. Out of the other two, the top one is the previous card that was already punched, and the bottom one is the next card to

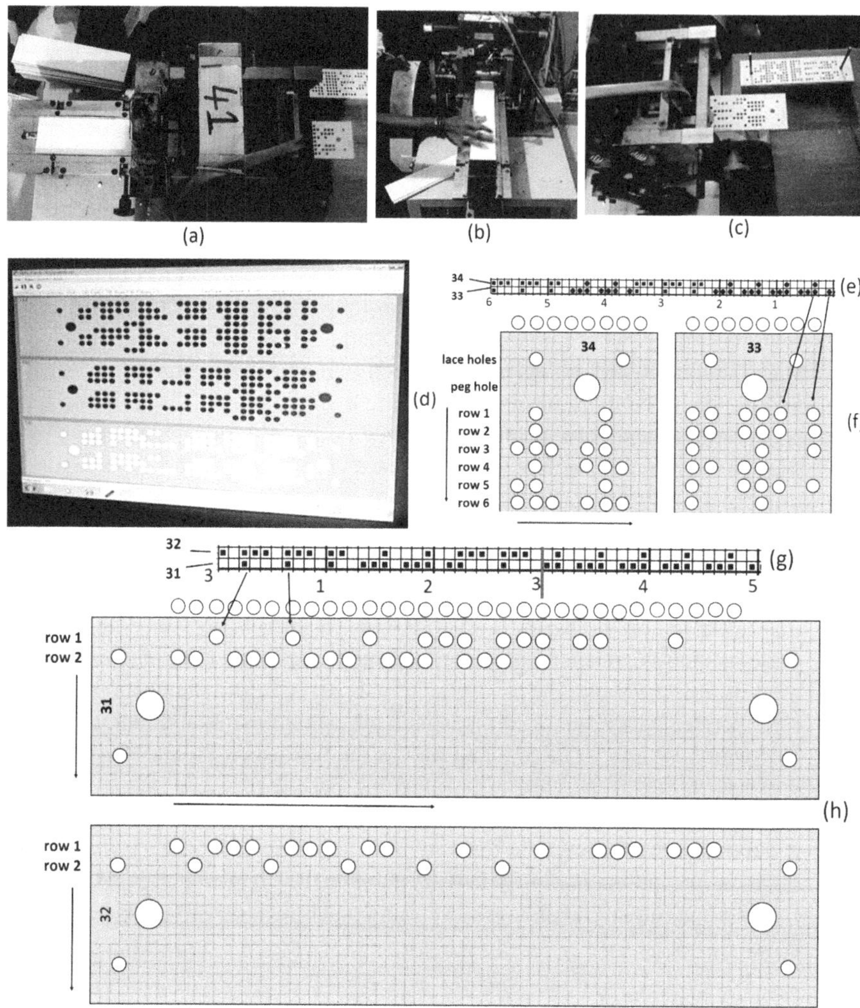

FIGURE 10.4 Electronic card punching machine and short-row-wise and long-row-wise punching: (a) bird's-eye view of the electronic card punching machine, (b) front view of the electronic card punching machine, (c) back view of the electronic card punching machine, (d) monitor attached with the electronic card punching machine displaying three cards, (e) two picks (33 and 34) of punching graph shown in Figure 10.3b, (f) two cards punched short row wise from the graph shown in (e), (g) two picks (31 and 32) of punching graph shown in Figure 10.3c, and (h) two cards punched long row wise from the graph shown in (g).

be punched. Thus, the monitor always informs the punching machine operator about the progress in the punching of the previous, present, and next cards. The punched card that came out of the machine is collected and arranged in the cardholder's nail as per the sequence. If any mistake in punching is noticed, the card is marked in red and kept separately. After completing one set of card punching, the wrongly punched cards are re-punched one by one by selecting the particular serial number of the pick. The clear perfection of the cutting edge of the holes is continuously monitored. If continuous imperfections are observed, the punches are replaced with new ones. The rate of punching is 60–80 cards per hour (240 hooks).

The top of Figure 10.4e shows the bottom two picks (33 and 34) of the 48 × 32 graph shown in Figure 10.3b prepared for short-row punching (8 ends per section). Figure 10.4f shows two cards punched by reading the graph in sets of eight-eight ends and punching the card short row wise by keeping the card vertically. The first card is punched as per the marks on the thirty-third pick, and the second card is punched as per the marks on the thirty-fourth pick. The reading of the ends on the graph is from right to left, and punching is also from right to left in each short row. These cards are punched using either the piano card punching machine or the electronic automatic card punching machine. Figure 10.4g shows the bottom two picks (31 and 32) of the 50 × 30 graph shown in Figure 10.3c prepared for long-row punching (30 ends per section). Figure 10.4h shows two cards punched by reading the graph in sets of thirty-thirty ends and punching the card long row wise by keeping the card horizontally. The first card is punched as per the marks in the thirty-first pick, and the second card is punched as per the marks in the thirty-second pick. The reading of the ends is from left to right, and punching is also from left to right in each row. These cards are punched using the manual punching plate.

Furthermore, the electronic punching machine is also used for long-row punching of cards. After loading the digital graph file in the processor for card punching, the monitor shows two options, viz., short-row punching or long-row punching. The designer can select either short-row punching or long-row punching. For both options, the card moves vertically (short row wise) in the machine. When the short-row punching option is selected, the processor reads the marks on sets of eight ends and transfers the signals for the corresponding punches to punch in the short row of the card. When the long-row punching option is selected, the processor reads the marks of the first end in each section of the thirty ends. That is, the processor reads the marks of 1st, 31st, 61st, 91st, 121st, 151st, 181st, and 211th ends—a total of eight ends—and transfers the signals to operate the keys and punch the first short row of the card. Next, the processor reads the marks of 2nd, 32nd, 62nd, 92nd, 122nd, 152nd, 182nd, and 212th ends—a total of eight ends—and transfers the signals to operate the keys and punch the second short row of the card. This will continue for 30 times for a 240-hook jacquard. For the 30th time, the processor reads the marks of 30th, 60th, 90th, 120th, 150th, 180th, 210th, and 240th ends—a total of eight ends—and transfers the signals to operate the keys and punch the last thirtieth short row of the card. This results in the card getting punched long row wise, even though the card is moving short row wise.

Apart from these, CAFGD is also used for hand plate punching. As already explained, in the hand punching method, there are two stages of punching: (i) marking over the card by reading the graph and (ii) punching by seeing the marks on the card. The CAFGD software is now very conveniently used for hand plate punching with long-row punching. After preparing the figured weave graph on the computer, all the picks of the graph are converted into the required form of the cards. The marks on the graph appear as circle hole marks in the card at the corresponding places, following a long-row-wise order on the card. The correct pitch space of the jacquard is also maintained in between the circle hole marks. The size of the card is marked with border outlines. These card layouts are taken as a paper printout. This paper printout is cut following the card outline, which becomes individual paper printouts of punched cards. This paper card printout is kept over the real card on the perforated punching plate. The round mark on the paper card guides the punching person to punch the holes very easily. This method avoids human error in reading the graph and saves a lot of time in hand punching plates.

10.4.3 MANUAL AND MACHINE CARD LACING

After completing the punching of all the cards, the cards are laced in sequential order and made in chain form. Two types of lacing are in practice: one is manual lacing, and the other is machine lacing. In manual lacing, people do the work of lacing by inserting lacing twine in the lacing holes of the card one after the other in crisscross order using lacing pins. In machine lacing, the machine does the lacing automatically. The operator of the machine has to feed the cards manually, one after the other, into the machine. Out of two lacing twines, one is fed from the top of the machine and the other is from the bottom, similar to the normal cloth stitching tailor machine. The needles in the machine do the lacing automatically.

Figure 10.5 shows the manual and machine lacing. Figure 10.5a shows how two people are doing the lacing work manually, sitting on the floor. One person is doing the lacing of the first lacing twine. Following the first person, the second person is doing the lacing work of the second lacing twine. Instead of sitting on the floor, a lacing stand can also be used for lacing. In the figure, it is also observed that the alternate cards are shaded with blue paint on the top. These are the plain cards punched in one-up, one-down order (long row wise). The unshaded cards are design cards. The blue shading on the plain card guides the handloom jacquard weaver to insert the plain pick when that card comes down from the cylinder after the selection of hooks. Figure 10.5b shows how the two thin lacing twine are crisscrossed to lace the card one after the other. One rope comes on top of the card (straight line) by crossing with the other rope at the bottom of the card. Figure 10.5c also shows how the two thin ropes are crisscrossed to lace the card one after the other. This crisscrossing differs little from the crisscrossing shown in Figure 10.5b. Figure 10.5d and e show the front and back-side views of the lacing machine. Figure 10.5d shows the operator placing the cards one after the other over the movable pegs. The rotating pegs carry the card to the needle lacing area, where the needle laces the card with the lacing twine passed through the needle. Figure 10.5e shows how the cards come out after lacing. After completing the lacing, the starting and ending lacing ropes are joined together to make the cards in continuous chain form. The card chain is mounted over the jacquard cylinder by making the peg holes in the card over the pegs in the cylinder.

Figure 10.5f shows the electronic jacquard mounted on a loom. Figure 10.5g shows the electronic design control box with a pen drive inserted on the right side. The digital figured graph design prepared (Figure 10.1e) is saved to this pen drive and transferred to the electronic control box. The pin on the top of this box is the connection to the electronic jacquard through which the data of digital graph design

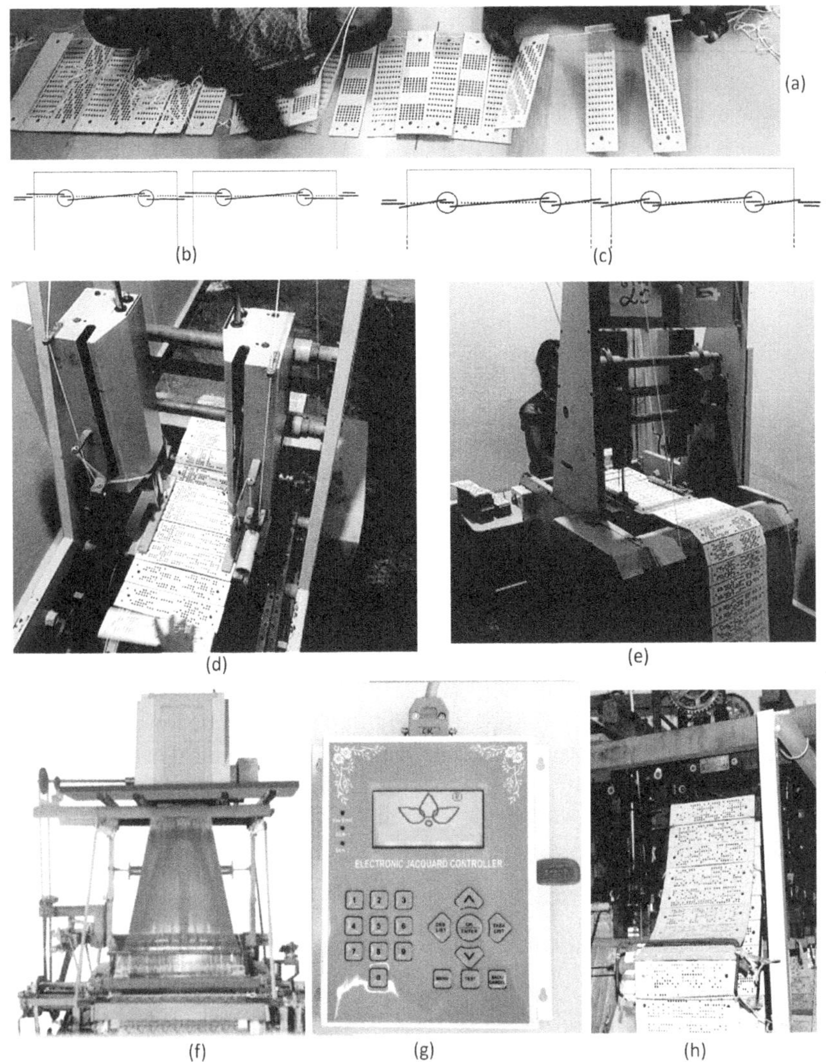

FIGURE 10.5 (a) Manual lacing of punched cards, (b) a method of crisscrossing the two lacing ropes for lacing the cords, (c) another method of crisscrossing the two lacing ropes for lacing the cords, (d) front view of the card lacing machine showing the card feeding, (e) back view of the card lacing machine showing the laced cards coming out of the machine, (f) a power loom mounted with electronic jacquard on the top of the loom, (g) an electronic digital figured design control box showing the pen drive on the right and transfer cable on the top, and (h) a chain of punched white cards mounted on the cylinder of mechanical jacquard.

gets transferred to the electronic jacquard. Figure 10.5h shows the chain of punched cards, white in colour, rolled over the cylinder of the mechanical jacquard mounted on a loom. The weaver takes care of the weaving and produces figured fabric on the loom. When a few repeats of the design are woven, the graph designer has to ascertain whether the figure and ground parts are as per the motif from which the graph design was made. The designer should also verify that the fabric is woven as per the structure of the weaves assigned and similar to the fabric simulation developed on the screen. The score of the designer depends upon how efficiently the designer has used the tools and options of computer-aided figured graph designing software for converting the figured motif into a perfect digital figured graph and the fabric simulation on the screen, which, when transferred to the loom, produces the actual figured fabric as per the desired end use.

10.5   ADVANTAGES OF CAFGD SOFTWARE

The following is the list of advantages of computer-aided figured graph designing software:

- Ease in enlarging the motif to the required number of ends and picks.
- Weave mapping is completed within no time for any graph size.
- Saving the graph file at each stage is possible. Many graphs can be prepared for one motif.
- Preparing the new motif directly on the screen is easy using the different drawing and editing tools.
- Different simulations and their printouts help to visualize the fabric before its actual weaving.
- Digital graph design is useful for electronic jacquard and electronic card punching and mechanical jacquard and manual-mechanical card punching.
- Transferring graph design and its simulations from one place to another through mail to get the concurrence and approval of concerned people is easy and fast.
- It enables error-free application of weaves on the graph and mostly error-free card punching.
- Storing the figured graphs and their retrieval are easy at any time.

10.6   CARING POINTS FOR THE FIGURED GRAPH DESIGNER

Following is the list of consolidated points that the figured graph designer should keep in mind and follow while using the CAFGD software:

- Note the quality parameters (ends per inch, picks per inch, jacquard capacity, and weaves) of the fabric for which the graph is to be prepared.
- Calculate the repeat width of the figure on the cloth and decide the motif length to be taken.
- Select or draw the figured motif to suit the calculated repeat width of the figure on the cloth.
- Calculate the proportionate motif length when the motif width differs from the repeat width of the figure on the cloth.
- Calculate the graph size to be prepared as per the quality parameters noted and the final calculated motif size.

- Scan the motif or cloth with the maximum possible resolution and colour, grey scale, or black-and-white colour modes as required.
- Reduce the graph size as required and reduce the colours to the maximum possible to edit the figure outlines easily.
- Do rescanning by changing the resolution and colours if the outlines of the figure parts are not clear for editing.
- Edit the outlines of the figure parts perfectly to enhance the shape of the figure parts.
- Ensure the repeating of the figure and ground parts with proper joining and continuing.
- Derive, create, and save the weaves to be applied on the different figure and ground parts.
- Use different colours for different weaves and note down which colour represents end up and which colour represents end down in each weave.
- Apply as many colours to different figure and ground parts of the edited motif as per the number of weaves to be applied, keeping in mind 'one colour for one weave'.
- Ascertain that the colours used for the weave application and the colours used in the edited motif are different.
- Contour the figure parts with one colour and the ground area with another colour.
- Select a suitable contour option to restrict or not to restrict the weave application in the corners and edge pixels as per the cloth variety to be produced.
- Check the end and pick floats thoroughly and bind the excess float length.
- Prepare the weave colour simulation by applying the required warp and weft colours.
- Prepare the fabric simulation by applying the actual interlacement of ends and picks with proposed yarn counts that are visible on the face side.
- Take the printouts of these simulations on paper and check their correctness.
- Drape the fabric simulation to ascertain the suitability of the figured graph prepared for the end use.
- Rework the graph from the beginning if the simulations do not match the desired end use.
- Convert and save the digital figured graph prepared as per the format required for weaving on the electronic jacquard.
- Convert and save the digital figured graph prepared as per the format required for punching the cards using the electronic card punching machine.
- Print the digital figured graph on paper with the required graph count and size for punching the cards using manual card punching plates or machines.
- Ascertain that the cards are punched without any mistake.
- Confirm that the digital figured graph prepared for the electronic jacquard produces the desired fabric without any mistakes.
- Verify that the punched cards for the mechanical jacquard produce the desired fabric without any mistakes.
- Make sure that the size and shape of the figure parts produced on the cloth are proportionate and perfect.

BIBLIOGRAPHY

1. ArahPaint. (2018). *User's Manual.* Arahne Software Company in Ljubljana, Slovenia.
2. Auto Tex 2000. (2005). *User's Manual.* PLC Consulting Company, New Delhi.
3. CadVantage Win Jacquard. (2001). *User's Manual.* Teckmen Systems Pvt. Ltd., Pune.
4. Grosicki, Z. J. (2004). Harness drawing in, card cutting, and card lacing. In *Watson's Textile Design and Colour* (pp. 188–189). Woodhead Publishing Limited, Cambridge.
5. Panneerselvam, R. G. (2013). Use of MS paint for jacquard graph designing and printing. *Indian Journal of Fibre & Textile Research*, 38, 186–192.
6. Textronics. (2000). *User's Manual of Design.* Jacquard Textronics Design Systems (I) Pvt. Ltd., Mumbai.

Index